MATHEMATICAL MODELLING

From Theory to Practice

SERIES ON MATHEMATICS EDUCATION

Published

Vol. 1 How Chinese Learn Mathematics
Perspectives from Insiders
Edited by: L. Fan, N.-Y. Wong, J. Cai and S. Li

Vol. 2 Mathematics Education
The Singapore Journey
Edited by: K. Y. Wong, P. Y. Lee, B. Kaur, P. Y. Foong and S. F. Ng

Vol. 4 Russian Mathematics Education
History and World Significance
Edited by: A. Karp and B. R. Vogeli

Vol. 5 Russian Mathematics Education
Programs and Practices
Edited by A. Karp and B. R. Vogeli

Vol. 7 Mathematics Education in Korea
Volume 1: Curricular and Teaching and Learning Practices
Edited by Jinho Kim, Inki Han, Joongkwoen Lee and Mangoo Park

Vol. 8 Mathematical Modelling
From Theory to Practice
Edited by N. H. Lee and D. K. E. Ng

Vol. 9 Primary Mathematics Standards for Pre-Service Teachers in Chile
A Resource Book for Teachers and Educators
By P. Felmer, R. Lewin, S. Martínez, C. Reyes, L. Varas, E. Chandía, P. Dartnell, A. López, C. Martínez, A. Mena, A. Ortíz, G. Schwarze and P. Zanocco

Vol. 10 Mathematics and Its Teaching in the Southern Americas
Edited by H. Rosario, P. Scott and B. R. Vogeli

Forthcoming

Vol. 3 Lesson Study
Challenges in Mathematics Education
Edited by M. Inprasitha, M. Isoda, B.-H. Yeap and P. Wang-Iverson

Vol. 6 How Chinese Teach Mathematics
Perspectives from Insiders
Edited by L. Fan, N.-Y. Wong, J. Cai and S. Li

Series on Mathematics Education Vol. 8

MATHEMATICAL MODELLING

From Theory to Practice

Edited by

Lee Ngan Hoe

Ng Kit Ee Dawn

National Institute of Education,
Nanyang Technological University, Singapore

World Scientific

NEW JERSEY • LONDON • SINGAPORE • BEIJING • SHANGHAI • HONG KONG • TAIPEI • CHENNAI

Published by

World Scientific Publishing Co. Pte. Ltd.
5 Toh Tuck Link, Singapore 596224
USA office: 27 Warren Street, Suite 401-402, Hackensack, NJ 07601
UK office: 57 Shelton Street, Covent Garden, London WC2H 9HE

Library of Congress Cataloging-in-Publication Data
Lee, Ngan Hoe.
Mathematical modelling : from theory to practice / Lee Ngan Hoe, National Institute of Education, Nanyang Technological University, Singapore, Ng Kit Ee Dawn, National Institute of Education, Nanyang Technological University, Singapore.
pages cm. -- (Series on mathematics education ; volume 8)
Includes bibliographical references.
ISBN 978-9814546911 (hardcover : alk. paper)
1. Mathematics--Study and teaching (Secondary) 2. Mathematics teachers--Training of. 3. High school teachers--Training of. I. Ng, Kit Ee Dawn. II. Title.
QA11.2.L445 2015
510.71'25957--dc23

2014045571

British Library Cataloguing-in-Publication Data
A catalogue record for this book is available from the British Library.

Typeset by Stallion Press
Email: enquiries@stallionpress.com

Printed in Singapore

AUTHORS' BIOGRAPHY

ANG Keng Cheng is currently an Associate Professor and the Head of the Mathematics and Mathematics Education academic group at the National Institute of Education, Singapore. He obtained his PhD in Applied Mathematics from the University of Adelaide, Australia. His research interests include computational methods in partial differential equations, numerical analysis, and the teaching and learning of mathematical modelling.

Raymond BROWN is an Associate Professor and the Deputy Head of School, Learning & Teaching, in the School of Education and Professional Studies at Griffith University, Australia. Recently, his research has been concerned with providing insights into how social processes interact at the individual and small group levels to motivate and guide students to 'speak' and 'act' as mathematicians when engaged in Mathematical Modelling.

CHAN Chun Ming Eric (PhD) is a lecturer with the Mathematics and Mathematics Education Academic Group, National Institute of Education, Singapore. He lectures on primary mathematics education in the pre-service and in-service programmes. He is the author of several mathematics resource books (primary levels) and is also the co-author of the Targeting Math series.

CHENG Lu Pien, received her PhD in Mathematics Education from the University of Georgia, United States in 2006. She is an Assistant Professor with the Mathematics and Mathematics Education Academic Group at the National Institute of Education, Singapore. She specialises in mathematics education courses for primary school teachers. Her research interests include the professional development of primary school mathematics teachers, tools and processes in mathematics education programmes for pre-service teachers, and children's thinking in the mathematics classrooms.

CHUA Kwee Gek was a teaching fellow and lecturer with the Mathematics and Mathematics Education Academic Group, National Institute of Education, Singapore. She has held the positions of Department Head and

Mathematics and Research and Evaluation Officer at the Singapore Ministry of Education. Currently, she is a part-time mathematics tutor and an external mentor for the Ministry's Improve Confidence and Achievement in Numeracy project.

Lyn D. ENGLISH is a Professor of mathematics education at the Queensland University of Technology. Her areas of research include mathematical modelling, mathematical reasoning and problem solving, engineering education, and statistical literacy. Lyn is a Fellow of The Academy of the Social Sciences in Australia and is founding Editor of the international journal, *Mathematical Thinking and Learning*.

Vince GEIGER is an Associate Professor in the Faculty of Education at Australian Catholic University, Australia. He is currently an Associate-Editor of the *Mathematics Education Research Journal*. Vince is also a former Secretary of the Mathematics Education Research Group of Australasia and a Past President of the Australian Association of Mathematics Teachers. His research interests include mathematical modelling, numeracy, the use of digital tools in mathematics teaching and learning, and the professional learning of mathematics teachers.

Susanne GRUENEWALD studied mathematics and history to become a teacher at the University of Hamburg, Germany. She holds a master degree in education. Currently she is completing her PhD at the University of Hamburg on empirical research in mathematical modelling.

HO Foo Him was a teaching fellow for the Mathematics and Mathematics Education Academic Group of the National Institute of Education, Singapore. His academic affiliations are as follows: MSc (Systems and Control), University of Sheffield, United Kingdom; BSc (Maths) National University of Singapore, Singapore; Diploma in Education, National Institute of Education, Singapore.

HO Siew Yin is currently a Post-Doctoral Fellow at the Faculty of Education, Science, Technology and Mathematics, University of Canberra, Australia. She holds a PhD in mathematics education from the National Institute of Education, Singapore. Her research interest is in visualisation and teacher education.

HO Weng Kin is currently an Assistant Professor at the Academic Group of Mathematics and Mathematics Education, National Institute of Education, Singapore. He holds a PhD in Computer Science from the University of Birmingham, United Kingdom. His research interest is in theoretical computer science, mathematics and mathematics education. For mathematics education, he is particularly interested in the use of history of mathematics and information technology in the teaching and learning of mathematics in secondary schools.

Toshikazu IKEDA is a Professor in mathematics education at the Faculty of Education and Human Sciences of Yokohama National University in Japan. He has been one of the members of the International Executive Committee of International Community of Teachers of Mathematical Modelling and Applications since 2005. Further, from 2010, he is one of the members of Mathematics Expert Group for PISA 2012.

JAGUTHSING Dindyal is currently an Associate Professor at the Mathematics and Mathematics Education Academic Group, National Institute of Education, Singapore. He holds a PhD in Mathematics Education from the Illinois State University, United States. His research interest is in various aspects of mathematics education including mathematical problem solving, international studies, teacher education and algebraic thinking.

Gabriele KAISER has been a Professor of mathematics education at the Faculty of Education of the University of Hamburg, Germany since 1998. Her areas of research include modelling and applications, international comparative studies, gender and cultural aspects in mathematics education as well as teacher education. Her latest project deals with teacher education under an international perspective. She is a prolific writer and since 2005 she has served as Editor-in-Chief of *ZDM - The International Journal on Mathematics Education*. She has served as the President of the International Community of Teachers of Mathematical Modelling and Applications. Currently she is convenor of the 13th International Congress on Mathematical Education.

Dawn LANG was Principal of A.B. Paterson College from 1994 to 2011. She is a member of the Board of Independent Schools Queensland and chairs the Block Grant Authority. Dawn organises and implements programmes for new principals as well as The Middle Managers as Leaders Programme for the Queensland Education Leadership Institute.

LEE Ngan Hoe is an Assistant Professor at the National Institute of Education, Singapore. He holds a PhD in Education from Nanyang Technological University, Singapore. His publication and research interests include the teaching and learning of mathematics at primary and secondary levels, covering areas such as mathematics curriculum development, metacognition and mathematical problem solving/modelling, technology and mathematics education, and textbooks and mathematics education.

NG Kit Ee Dawn is a lecturer in the Mathematics and Mathematics Education Academic Group at the National Institute of Education, Singapore. She holds a PhD in mathematics education from The University of Melbourne, Australia. She is involved in teacher professional development at both primary and secondary levels. Her research interests include the use of contextualised tasks in the teaching and learning of mathematics and fostering students' mathematical thinking and metacognition.

Trevor REDMOND has been a learner and teacher of mathematics for many years. He uses mathematical modelling to develop in students an interest of making mathematics through making it relevant to them. He is interested in seeing students develop their mathematical understanding and thinking. He conceptualised and has coordinated the A.B. Paterson College Mathematical Modelling Forum and Challenge from its inception. Trevor is now the Head of Mathematics at Somerville House, a girls' school in Brisbane, Australia.

Joanne SHEEHY has been the Head of mathematics for the past decade and an advocate, supporter and mentor of the A. B. Paterson College Mathematical Modelling Forum and Challenge. She sees that having an understanding of mathematical modelling assists students to generate resilience in solving problems, allows them to access problems and become better thinkers.

SOON Wan Mei Amanda is an Assistant Professor in the Mathematics and Mathematics Education Academic Group at the National Institute of Education in Singapore. She received her PhD in mathematics from the National University of Singapore. Her main research interests include the pricing of multiple products and the teaching and learning of mathematical modelling, in particular with differential equations.

Gloria STILLMAN is an Associate Professor at the Australian Catholic University, Australia. Currently, she holds two positions as the Deputy Head of Education in the Faculty of Education and Arts (Victoria) as well as the Head of Education in the Faculty of Education and Arts (Ballarat) at her university. She has a PhD in mathematics education from the University of Queensland, Australia. Her current major research interest is in teaching and assessing mathematical modelling in secondary schools. She is a prolific writer who has published extensively in areas such as metacognition, applications and modelling, curriculum development and teacher education. She is now the President of the International Community of Teachers of Mathematical Modelling and Applications.

YEO Kai Kow Joseph is a faculty member in the Mathematics and Mathematics Education Academic Group at the National Institute of Education, Singapore. As a teacher educator, he is involved in training pre-service and in-service mathematics teachers at primary and secondary levels and has also conducted numerous professional development courses for teachers in Singapore and overseas. Before joining the National Institute of Education in 2000, he held the post of Vice Principal and Head of Mathematics Department in secondary schools. His research interests include mathematical problem solving in the primary and secondary levels, mathematics pedagogical content knowledge of teachers, mathematics teaching in primary schools and mathematics anxiety.

ACKNOWLEDGEMENTS

Jonas Bergman ÄRLEBÄCK
Linköping University, Sweden

Jill BROWN
Australian Catholic University, Australia

Lyn ENGLISH
Queensland University of Technology, Australia

Peter GALBRAITH
University of Queensland, Australia

Toshikazu IKEDA
Yokohama National University, Japan

LEE Peng Yee
National Institute of Education, Nanyang Technological University, Singapore

Mogens NISS
Roskilde University, Denmark

NG Kit Ee Dawn
National Institute of Education, Nanyang Technological University, Singapore

Gloria STILLMAN
Australian Catholic University, Australia

CONTENTS

CHAPTER 1

INTRODUCTION: MATHEMATICAL MODELLING OUTREACH IN SINGAPORE

NG Kit Ee Dawn

LEE Ngan Hoe

This book documents the journey undertaken by educators from the Mathematics and Mathematics Education (MME) Academic Group in the National Institute of Education (NIE) and Singapore schools during a Mathematical Modelling Outreach (MMO) event in June 2010 under the guidance of renowned experts in the field of mathematical modelling. In this introductory chapter, we provide the context in which MMO was organised, the theoretical framework of mathematical modelling adopted for MMO, and outline the focuses of mathematical modelling during MMO in relation to the Singapore mathematics curriculum framework. We will also present an overview of the chapters in this book, summarise the progress of modelling efforts in schools to date, and draw implications for future directions on mathematical modelling in schools.

Keywords: Facilitation, Mathematical Modelling Outreach, Modelling Cycle, Singapore Schools, Teacher Education.

1.1. Mathematical Modelling Outreach (MMO)

According to the Organisation for Economic Co-Operation and Development (OECD), mathematical literacy refers to "an individual's capacity to formulate, employ, and interpret mathematics in a variety of contexts" (OCED, 2013, p. 5). Mathematical literacy also includes the ability to reason mathematically and to make appropriate choices in use of mathematical concepts, skills, and tools to describe, explain and predict real-world situations. In other words, mathematical literacy is perceived as a part of and within the real-world as an individual harnesses the mathematics he or she possesses for decision making.

Although for many years, Singapore students have been top scorers in mathematics for international studies such as Trends in International Mathematics and Science Study (TIMSS) and Programme for International Student Assessment (PISA) (Ministry of Education [MOE], 2011; 2014a), educators in Singapore recognise the importance of preparing students towards mathematical literacy along with equipping students with 21st century competencies (MOE, 2014b). Mathematical modelling is one of the ways towards this goal.

Proponents of mathematical modelling have argued that the mathematical modelling promote connections between school-based mathematics and real-world problem solving (Blum, 2002; English, 2009; Kaiser & Maaß, 2007; Lesh & Doerr, 2003; Niss, 2010; Stillman, Brown, & Galbraith, 2008). In particular, modelling eliciting activities involving complex data embedded within real-world interdisciplinary situations have been reported as rich platforms to inculcate mathematical reasoning and sense making, making learning more meaningful to younger children (English, 2009).

1.1.1. *Context for MMO in Singapore*

Mathematical modelling has become the foundation for the PISA 2015 mathematics framework (OECD, 2013) and it has been incorporated into the curricula of education systems all over the world. This is seen from the growing participation of the biennial International Community of Teachers of Mathematical Modelling and Applications (ICTMA) conferences which showcase representative research, teaching, and learning on modelling from different continents. For example, in Germany, Professor Gabriele Kaiser and her colleagues from The University of Hamburg have been organising modelling weeks for upper and lower secondary students for more than a decade. During these modelling weeks, students work in small groups guided by their mentors (e.g., student-teachers and educators from the university) to solve real-world problems. Chapter 2 of this book provides an excellent description of the modelling week. In Australia, Associate Professor Raymond Brown from the Griffith University at the Gold Coast has been collaborating with Associate Professors Gloria Stillman and

Vincent Geiger from the Australian Catholic University along with other mathematics educators such as Mr Trevor Redmond previously from A. B. Paterson College to facilitate Mathematical Modelling Challenges for many years. Modelling Challenges involves primary, secondary, and pre-university students working in groups on a range of modelling activities in week-long infusion events under the tutelage of experts in the field of mathematical modelling. Representatives from Singapore secondary schools have begun to participate in Modelling Challenges regularly since 2009. Australian schools in Queensland have long infused mathematical modelling in their curriculum. Chapters 5 and 9 in this book provide readers with insights into teaching and learning with mathematical modelling in the Australian context.

In Singapore, mathematical modelling was first incorporated in the mathematics curriculum framework in 2003. However, the introduction to mathematical modelling was subtle, with limited teacher education on the value of modelling activities. Schools were free to interpret the learning outcomes from modelling activities which were often viewed as optional enrichment tasks. Singapore teachers and students face constraints in time management due to traditional high stakes assessment systems. This in turn reduces the options to various mathematical learning activities. If modelling tasks were not integrated into the mainstream curriculum focuses in teaching, learning, and assessment, it would be challenging for Singapore teachers to harness the potentials of such tasks (Ang, 2010a). Furthermore, Singapore teachers were mostly unfamiliar with what mathematical modelling is, with many associating it with drawings of bar models commonly used for problem solving in schools (Chan, 2013). Many teachers were also unable to distinguish between applications and modelling as both refer to problem solving in real-world contexts (Ng, 2013a). Stillman, Brown, and Galbraith (2008) provided an effective way differentiating between applications and modelling. Application tasks are adopted when the teacher already has in mind specific taught mathematical knowledge and skills to be used and searches for appropriate real-world situations to showcase the application. On the other hand, mathematical modelling starts with the real-world context which draws upon different mathematical knowledge and skills depending on the interpretations of the modellers during model development to represent the problem solving context. Whilst application tasks have been used frequently in Singapore schools for many years, modelling was still a predominantly new idea which requires a mind set change in teachers' conceptions of the problem solving process as well as pedagogical focuses (Ng, 2010). The open-ended non-linear problem solving process during mathematical modelling poses varying challenges to Singapore teachers and their students. Teachers generally lack awareness of the facilitation focuses during the cyclical modelling stages, specifically how students can be guided in their mathematisation processes between these stages. With reports on the impact of modelling activities on students' engagement with mathematics few and far between, mathematical modelling has remained quite a mystery to Singapore schools.

The lack of teacher readiness in mathematical modelling both in mind set and facilitation repertoire needed to be addressed (Ng, 2010; 2013b). In a concerted push towards the infusion of mathematical modelling in schools, a range of efforts from the Mathematics and Mathematics Education Academic Group (MME) in the National Institute of Education (NIE) and Singapore Ministry of Education started in 2009 (Ang, 2013; Balakrishnan, Yen, & Goh, 2010, Balakrishnan, 2011; Curriculum Development and Planning Division [CPDD], 2009; Lee, 2013; Ng & Lee, 2010; Tan & Ang, 2013). First, opportunities needed to be provided for primary and secondary school teachers to experience mathematical modelling lessons and witness students' mathematisation processes during model development. This was the initial impetus for Singapore's first Mathematical Modelling Outreach (MMO) event organised by MME at NIE Singapore in June 2010. Associate Professor Ang Keng Cheng, a pioneer researcher on mathematical modelling in Singapore, as well as modelling experts in Australia and Germany became pillars of support for MMO.

1.1.2. *Objectives and structure of MMO*

MMO was set to achieve three main objectives. Its main goal was to reach out to Singapore primary and secondary schools and introduce the potentials of mathematical modelling as a platform for eliciting mathematical thinking, communication, and reasoning among students. Another purpose was to make connections between school-based mathematics and real-world problem solving more explicit for students through model development. It is hoped that students would build upon their existing mathematics repertoire whilst creating mathematical arguments for specific real-world contexts in the modelling process. A third objective of MMO was to provide teachers with preliminary teacher education on (a) what mathematical modelling is, (b) how to facilitate the modelling process, and (c) task design guidelines for crafting modelling problems.

MMO was designed to be a three-day immersion programme for school representatives from 1st to 3rd June 2010. A total of 13 and 15 Singapore primary and secondary schools respectively participated in MMO along with one primary school from Bandung, Indonesia, and one secondary school from the Gold Coast, Australia. Each school sent a team of one teacher and four students for MMO. A two-pronged approach was adopted during MMO. Firstly, students formed inter-school working groups of four at primary and secondary levels to solve modelling problems facilitated by NIE pre-service teachers from the third year of Bachelor of Arts and Science Degree programme. The pre-service teachers were selected based on their mathematics results and teaching performance. They worked in pairs and each pair is placed under the tutelage of a mathematics educator from MME to prepare for the facilitation of an assigned modelling problem. The pre-service teachers and their mentors underwent several sessions of mathematical modelling problem solving and facilitation process conducted by the first author before they prepared for their assigned modelling task in MMO. Student-groups from schools were provided with scaffolding worksheets and resources to complete their modelling tasks over the first two days of MMO. They spent the third day preparing for their sharing and giving verbal presentations of their journey in developing the mathematical model for the given real-world problem. Feedback on the mathematical models created by student-groups was provided by the pre-service teachers and their MME mentors. Secondly, the accompanying teachers for MMO were invited to participate in teacher workshops on mathematical modelling with respect to (a) to (c) above as well as seminars presented by invited speakers. Details about the teacher workshops can be found in Chan (2013), Lee (2013), and Ng (2013a). The workshops and seminars were planned with a view of sharing ideas on how mathematical modelling can be infused in Singapore mathematics classrooms. Representatives from the Singapore Ministry of Education who were crafting guidelines to schools on mathematical modelling then also contributed to the question-and-answer sessions in each workshop. In addition, the teachers were also invited to sit in during some of the facilitation sessions conducted by the pre-service teachers so that they could experience first-hand how a modelling activity can be conducted. Student work from MMO were presented on 4th June 2010 during the second Lee Peng Yee Symposium where renowned modelling experts were also invited as keynote speakers to provide more holistic views about what modelling can be interpreted as at the primary and secondary levels in different education systems around the world and offer practical suggestions for teachers on how to infuse modelling tasks in their day-to-day teaching activities.

1.2. Theoretical Framework for Mathematical Modelling in MMO

Mathematical modelling is referred to as a process of representing real-world problems in mathematical terms in an attempt to understand and find solutions to the problems (Ang, 2010b). Blum and Niss (1991) had considered this as the "applied problem solving process" (p. 38). There have been many diagrammatic representations of this applied problem process or mathematical modelling process discussed in modelling research throughout the years. Chapter 2 in this book by Kaiser and Grünewald provides a comprehensive but critical history of the conceptual frameworks underlying the myriad diagrams. In this section, we cite de Lange (2006) as the starting platform for our theoretical framework in MMO. There are parallel concepts between this framework and those used in PISA's mathematics framework (OECD, 2013) and the current Singapore mathematics syllabus (CPDD, 2012a).

Figure 1.1 is a simplistic adaptation of de Lange's conceptual framework of the mathematical modelling process which he referred to as the "mathematisation cycle" (de Lange, 2006, p. 17). de Lange emphasised that mathematisation is key to mathematical literacy which involves the appropriate and flexible use of mathematics within real-world contexts. Hence, mathematisation process always begins from an applied problem or the ***real-world problem***. This has been the case for most diagrammatic representations of the mathematical modelling process. The real-world problem is then

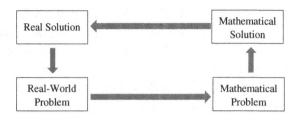

Figure 1.1. An adaptation of the mathematisation cycle from de Lange (2006, p. 17).

analysed using mathematical lenses. Here, the modeller selects appropriate mathematical concepts, skills, and tools based on his or her interpretation of the context the real-world problem is situated in. This results in a ***mathematical problem*** which is essentially a simplified version of the actual real-world problem with assumptions made, conditions identified, and key variables quantified mathematically. de Lange (2006) highlighted the gradual "trimming away" (p. 18) of reality when formulating the mathematical problem as a necessary procedure but cautioned that the mathematical problem should represent the real-world problem faithfully. A ***mathematical solution*** follows where the modeller attempts to bring fruition to the preliminary mathematical model. The modeller later tries to make sense of the solution in terms of the real-world context and examines the solution for its actual applicability in reality. A ***real-world solution*** develops from the preliminary model. In essence, the process of mathematisation is cyclical because the modeller has to move between the four components in Figure 1.1 when reviewing the limitations of the initial mathematical model and refining it.

1.2.1. *Theoretical framework of modelling MMO in relation to the Singapore mathematics curriculum*

In the design and selection of tasks for MMO, we adopted two theoretical perspectives of mathematical modelling in educational contexts outlined by Kaiser and Sriraman (2006): realistic modelling and contextual modelling. Realistic modelling serves pragmatic-utilitarian goals which promote the solving of real-world problems to support the relevance of school mathematics and to nurture modelling competencies (e.g., Kaiser, Maaß, 2007). Contextual modelling views mathematical models as "purposeful conceptual systems" (Lesh, 2003, p. 44) which describes, explains, or predicts real-world phenomena. Proponents of this perspective (e.g., English, 2009; Lesh & Doerr, 2003) emphasise the mathematisation of a situation in which modellers devise or select appropriate symbolic mathematical representations (i.e., model development) for the given real-world situation in meaningful ways (e.g., explaining, justifying, predicting, conjecturing, representing, and quantifying). An important goal of the contextual modelling perspective is for modellers to develop, critique, and validate their own mathematical models in view of the assumptions and conditions set within real-world contexts.

In the recent syllabus documents, the Singapore Ministry of Education articulated the mathematical modelling process explicitly and provided examples of modelling tasks (CPDD, 2012b). The key components in the theoretical framework adopted by MMO described above shared crucial features with the mathematical modelling process espoused by the syllabus documents. However, the syllabus highlighted four "elements" of the modelling process (i.e., formulating, solving, interpreting, and reflecting) which depict the transition between the four components of Figure 1.1. For example, the element of formulating occurs when the modeller makes assumptions to simplify the real-world problem so as to form the mathematical problem. It was evident from the syllabus documents that plans were underway to equip Singapore students with modelling competencies and at the same time provide them with opportunities to use mathematics to describe, explain, and even predict real-world situations.

1.2.2. *Facilitation framework during MMO*

The facilitation framework implemented by pre-service teachers during student-group discussions in MMO was elaborated in detail in Ng (2013a). It comprised six phases extended from Polya's four stages (1957) of problem solving process: (a) understand the problem, (b) plan, (c) implement plan, and (d) reflect. Singapore mathematics teachers have been introduced to Polya's work as part of their pre-service education. Focused questions are aligned to each of the six phases to develop students' modelling competencies. These phases can be mapped to the four components of the modelling process articulated in Figure 1.1.

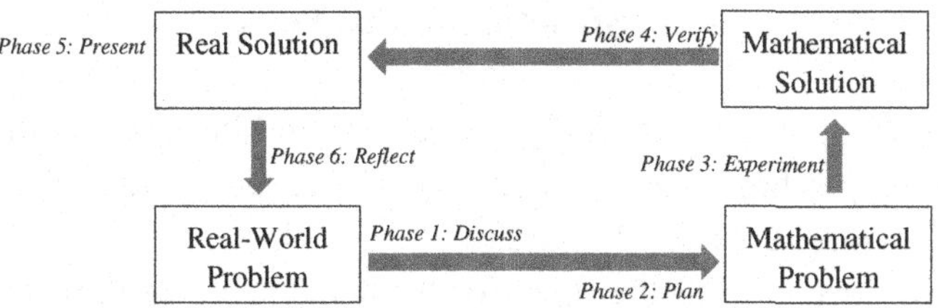

Figure 1.2. Six phases in the facilitation framework embedded in the modelling process.

Figure 1.2 shows the facilitation framework embedded within the modelling process. In Phase 1 (Discuss), student-groups are expected to come to a consensus when posing the mathematical problem from the given real-world problem through setting goals, making assumptions, identifying parameters, and defining key variables. In Phase 2 (Plan), careful planning (e.g., collection of relevant information) and organisation of approaches (e.g., use of certain ICT tools) to solve the mathematical problem posed is required. Phase 3 (Experiment) is crucial as it guides student-groups to carry out their plan and subsequently make appropriate conjectures throughout their solution process. Here, they test analysis the information they have, test conjectures, and begin developing a preliminary model to represent the real-world problem mathematically. A solution to the mathematical problem is sought at this phase. In this phase, student-groups are also prompted to review their initial plans and perhaps make adjustments to revise their initial mathematical models according to the results of their tested conjectures. Phase 4 (Verify) guides students to check the accuracy of their calculations, the reasonableness of their approaches, and the interpretations of their subsequent conclusions based on the real-world situation. This phase encourages student-groups to generate a real-world solution from the mathematical solution for more meaningful sense making. Phases 1 to 4 are cyclical together with the four components of the modelling process. By Phase 5 (Present), student-groups attempt to finalise their models and craft responses to the real-world problem, providing mathematical justification for their responses. Finally, in Phase 6 (Reflect), students review the validity of their models in relation to the real-world problem and discuss the appropriateness of their attempts in addressing the problem. They are also requested to propose possible applications of their models in other real-world situations.

1.3. Structure of this Book

This book documents the journey educators from MME took with primary and secondary school teachers and students during MMO. The book is organised in three connected sections.

The first section sets the stage for mathematical modelling in schools by presenting how mathematical modelling is infused in curricula at different parts of the world: Germany, Australia, Japan, and Singapore. Chapter 2 of this book by Kaiser and Grünewald examines the theoretical perspectives of mathematical modelling, its historical development, and the influence of mathematical modelling around the world before presenting how modelling competencies are promoted among school students participating in modelling weeks at the University of Hamburg. Together, Chapters 3 and 5 by Stillman and Geiger respectively outline how mathematical modelling has been infused in the Australian curriculum. Of particular interest is Stillman's chapter which introduces the differences between problem finding and problem posing for mathematical modelling. This chapter exemplifies the theory-practice nexus of mathematical modelling and brings readers to understanding that mathematical modelling is about how we perceive our world. Geiger's chapter provides readers with a glimpse of how mathematical modelling is embedded in the Australian Curriculum and discusses the role of technology in modelling tasks. Chapter 6 by Ikeda offers a candid view of the challenges faced by Japanese educators in incorporating modelling in schools and argues for the role open-ended real-world problems such as modelling problems within the Japanese curriculum. Lastly, Chapter 4 by Ang examines the potentials of mathematical modelling in Singapore schools and proposes a teacher education framework for the instruction of mathematical modelling in schools.

The second section of this book explores how mathematical modelling can be fostered in schools. Chapter 7 by English offers useful suggestions on how model-eliciting tasks can be adopted at primary levels for young children to learn mathematical ideas. Chapter 8 by Stillman very aptly discusses the issues associated with changes in practice, knowledge, and beliefs when incorporating mathematical modelling in schools in the context of the Queensland senior secondary mathematics curriculum in Australia. These echo the challenges Singapore is currently facing with the introduction of mathematical modelling in the Singapore curriculum (see Ang, 2010a; Ng, 2013b). The chapter lends weight to the educational purposes of modelling and suggests the use of metacognitive monitoring by teachers in facilitating the modelling process. It argues for teacher reflection on students' metacognitive activities on both the macroscopic (i.e., across the modelling process) and microscopic (i.e., within each component of the modelling process) levels during facilitation. The chapter also outlines pragmatic practices to promote reflective learning by students and proposes key focuses when evaluating students' modelling attempts. Chapter 9 by Brown, Redmond, Sheehy and Lang examines the role of collective argumentation, a socio-cultural approach to teaching and learning, in facilitating students' engagement in mathematical modelling. The chapter illustrates how this view of facilitation is manifested in student-group activities during an inter-school mathematical modelling challenge.

Finally, the third section of this book reports the nature of students' modelling process and teacher facilitation during the seven tasks used in MMO. Chapters 10 to 16 by MME educators serve as exemplars for the potentials of mathematical modelling in Singapore schools. These chapters examine the mathematisation processes of students during the task and draw implications on teaching and learning with modelling activities. Details of the tasks and scaffolding materials can be found in Ng and Lee (2012).

1.4. Moving Forward from MMO: The Ripple Effect Continues

MMO has provided a leading platform for initial dialogues between curriculum planners, mathematics educators, teachers, and students in exploring the potentials of modelling activities and implementation needs. The success of MMO could be viewed on three fronts in relation to its objectives. Firstly, MMO had served as a successful springboard enhancing further collaborative teacher education efforts between MME and the Singapore Ministry of Education towards a more concerted plan for the incorporation of mathematical modelling in schools. Secondly, as documented in the third section of this book, students' mathematical models during the tasks used in MMO showed explicit connections between school-based mathematics and real-world interpretations of the problem situation. In these chapters, MME educators had discussed and analysed the mathematisation process of each task, helping teachers and interested readers visualise the potentials of mathematical modelling in the context of Singapore schools. Thirdly, data from teacher workshops during MMO (Chan, 2013; Lee, 2013; Ng, 2013a) presented preliminary insights into the conceptions of mathematical modelling by teachers, the challenges teachers faced in problem posing and task design, as well as teachers' needs in subsequent professional development on the facilitation of modelling activities.

Since MMO, interest in mathematical modelling has gained momentum in schools, observed from increasing participation in modelling challenges overseas and sustained annual local inter-school modelling events (Kwek & Ko, 2011). Meanwhile, curriculum documents (CPDD, 2012b) outlined the key focuses of mathematical modelling in Singapore schools and included explicit modelling examples for various levels along with facilitation questions for each stage of the modelling process. Addressing some of the teachers' needs identified during MMO, MME educators added to the resources available for Singapore teachers through a task book (Ng & Lee, 2012) which contained task sheets, teacher facilitation prompts, and model evaluation rubrics. In-service courses for mathematical modelling at secondary levels conducted by MME educators and the Singapore Ministry of Education were put in place to provide teachers with facilitation suggestions for mathematisation during the modelling cycle, task design guidelines, and model evaluation focuses. Furthermore, schools also organised their own workshops for teachers when they incorporated mathematical modelling activities. Singapore research projects in mathematical modelling at the primary (e.g., Chan, Ng, Widjaja, & Seto, 2012) and secondary (e.g., Tan & Ang, 2013) levels are being conducted to explore teacher education issues in mathematical modelling so as to chart future in-service courses. Mathematics educators continue to raise awareness of the potentials of mathematical modelling among schools through symposia (see Galbraith, 2013) at conferences.

However, the journey into the world of mathematical modelling for Singapore schools has only just begun. The pebble has been thrown into the pond in 2010 and we have reported the initial ripple effect above, substantiated by various documentations which followed MMO. For this ripple effect to endure the challenges of time, mathematics educators would need to continue building on factors for success in the incorporation of mathematical modelling in Singapore schools and thus continue to fill in the gaps in teaching, learning, and research. A factor for success identified by Stillman (2010) was teacher readiness. Teacher readiness can be viewed from many perspectives. One of which was explored by Ng (2013b) which examined the importance of teachers as modellers first before they can harness the possibilities of teaching and learning offered by mathematical modelling. Another perspective is more holistic, looking into the role of assessment in mediation with curriculum initiatives. Educators may consider how the evaluation of learning during mathematical modelling could assimilate into the existing assessment climate of Singapore schools in order to develop teacher readiness for such activities.

This book contributes to the expanding literature on mathematical modelling by offering voices from the Singaporean context. It suggests how theoretical perspectives on mathematical modelling can be transformed into actual practice in schools, all within the existing infrastructure of the current Singapore mathematics curriculum. Moreover, the book provides documentary evidence on how plans put in place through MMO in 2010 have since been realised. Hence, the publication of this book is timely at this juncture. Not only does the book record how MMO was among the first pebbles launched into the pond. It also serves as a bridge over which educators can stand upon to view how the ripple effect had developed from the initial MMO pebble and the directions it may continue to extend. Perhaps in the process, other ripples in the teaching, learning, and research of mathematical modelling can be created.

References

Ang, K. C. (2010a). Mathematical modelling in the Singapore curriculum: Opportunities and challenges. In A. Araújo, A. Fernandes, A. Azevedo & J. F. Rodrigues (Eds.), *Educational interfaces between mathematics and industry: Proceedings of the EIMI 2010 Conference* (pp. 53–62). New York: Springer.

Ang, K. C. (2010b, December 17–21). *Teaching and learning mathematical modelling with technology.* Paper presented at the Linking applications with mathematics and technology: The 15th Asian Technology Conference in Mathematics, Le Meridien Kuala Lumpur.

Ang, K. C. (2013). Real-life modelling within a traditional curriculum: Lessons from a Singapore experience. In G. Stillman, G. Kaiser, W. Blum & J. Brown (Eds.), *Teaching mathematical modelling: Connecting to research and practice* (pp. 131–140). Dordrecht: Springer.

Balakrishnan, G., Yen, Y. P., & Goh, L. E. E. (2010). Mathematical modelling in the Singapore secondary school mathematics curriculum. In B. Kaur & J. Dindyal (Eds.), *Mathematical applications and modelling: Yearbook 2010* (1st ed., pp. 247–257). Singapore: Association of Mathematics Educators.

Balakrishnan, G. (2011). *Mathematical modelling: Insights from Singapore*. Paper presented at the Connecting to practice — Teaching practice and the practice of applied mathematicians: The 15th International Conference on the Teaching of Mathematical Modelling and Applications, Australian Catholic University (St. Patrick), Melbourne, Australia.

Blum, W., & Niss, M. (1991). Applied mathematical problem solving, modelling, applications, and links to other subjects: State, trends and issues in mathematics instruction. *Educational Studies in Mathematics*, 22(1), 37–68.

Blum, W. (2002). ICMI Study 14: Applications and modelling in mathematics education. *Educational Studies in Mathematics*, 51(1/2), 149–171.

Chan, C. M. E. (2013). Initial perspectives of teacher professional development on mathematical modelling in Singapore: Conceptions of mathematical modelling. In G. Stillman, G. Kaiser, W. Blum & J. Brown (Eds.), *Teaching mathematical modelling: Connecting to research and practice* (pp. 405–414). Dordrecht: Springer.

Chan, C. M. E., Ng, K. E. D., Widjaja, W., & Seto, C. (2012). Assessment of primary 5 students' mathematical modelling competencies. *Journal of Science and Mathematics Education in Southeast Asia*, 35(2), 146–178.

Curriculum Planning and Development Division [CPDD]. (2009). *Mathematics modelling in our mathematics curriculum*. Briefing for Heads of Department in Mathematics (Secondary Schools), at Unity Secondary School. Singapore: Ministry of Education.

Curriculum Planning and Development Division [CPDD]. (2012a). *O-Level mathematics teaching and learning syllabus*. Singapore: Ministry of Education.

Curriculum Planning and Development Division [CPDD]. (2012b). *Mathematical modelling resource kit*. Singapore, Ministry of Education: Author.

de Lange, J. (2006). Mathematical literacy for living from OECD-PISA perspective. *Tsukuba Journal of Educational Study in Mathematics*, 25(1), 13–35.

English, L. (2009). Promoting interdisciplinarity through mathematical modelling. *ZDM — The International Journal on Mathematics Education*, 41(1–2), 161–181.

Galbraith, P. (2013). *Discussant notes for symposium on mathematical modelling*. Paper presented at the Thinking: Time for a rethink? The 5th International Redesigning Pedagogy Conference National Institute of Education, Singapore.

Kaiser, G., & Maaß, K. (2007). Modelling in lower secondary mathematics classroom — problems and opportunities. In W. Blum, P. Galbraith, H. W. Henn & M. Niss (Eds.), *Modelling and applications in mathematics education: The 14th ICMI study* (pp. 99–108). New York: Springer.

Kaiser, G., & Sriraman, B. (2006). A global survey of international perspectives on modelling in mathematics education. *ZDM — The International Journal of Mathematics Education*, 38(3), 302–310.

Kwek, M. L., & Ko, H. C. (2011). *The teaching and learning of mathematical modelling in a secondary school.* Paper presented at the The 15th International Conference on the Teaching of Mathematical Modelling and Applications: Connecting to practice — Teaching practice and the practice of applied mathematicians Australian Catholic University (St. Patrick), Melbourne, Australia.

Lee, N. H. (2013). Initial perspectives of teacher professional development on mathematical modelling in Singapore: Problem posing and task design. In G. Stillman, G. Kaiser, W. Blum & J. Brown (Eds.), *Teaching mathematical modelling: Connecting to research and practice* (pp. 415–426). Dordrecht: Springer.

Lesh, R. A. (2003). How mathematizing reality is different from realising mathematics. In S. J. Lamon , W. A. Parker & K. Houston (Eds.), *Mathematical modelling: A way of life — ICTMA 11* (pp. 38–52). Chichester, UK: Horwood.

Lesh, R., & Doerr, H. M. (2003). Foundations of a models and modeling perspective on mathematics teaching, learning, and problem solving. In R. Lesh & H. M. Doerr (Eds.), *Beyond constructivism: Models and modeling perspectives on mathematics problem solving, learning, and teaching* (pp. 3–33). Mahwah, NJ: Lawrence Erlbaum Associates.

Ministry of Education [MOE]. (2011, 11 December). Press release: International studies affirm Singapore students' strengths In reading, mathematics & science Retrieved December 12, 2013, from http://www.moe.gov.sg/media/press/2012/12/international-studies-affirm-s.php.

Ministry of Education [MOE]. (2014a, 1 April). Singapore students excel in thinking flexibility And creatively to solve complex and unfamiliar problems, says PISA study Retrieved April 10, 2014, from http://www.moe.gov.sg/media/press/2014/04/singapore-students-excel-in-thinking-flexibility.php

Ministry of Education [MOE]. (2014b, April 1). Information sheet on 21st century competencies Retrieved April 10, 2014, from http://www.moe.gov.sg/media/press/2014/04/information-sheet-on-21st-century.php.

Ng, K. E. D. (2010). Initial experiences of primary school teachers with mathematical modelling. In B. Kaur & J. Dindyal (Eds.), *Mathematical modelling and applications: Yearbook of Association of Mathematics Educators* (pp. 129–144). Singapore: World Scientific.

Ng, K. E. D. (2013a). Initial perspectives of teacher professional development on mathematical modelling in Singapore: A framework for facilitation. In G. Stillman, G. Kaiser, W. Blum & J. Brown (Eds.), *Teaching mathematical modelling: Connecting to research and practice* (pp. 427–436). Dordrecht: Springer.

Ng, K. E. D. (2013b). Teacher readiness in mathematical modelling: Are there differences between pre-service and experienced teachers? In G. Stillman, G. Kaiser, W. Blum & J. Brown (Eds.), *Teaching mathematical modelling: Connecting to research and practice* (pp. 339–348). Dordrecht: Springer.

Ng, K. E. D., & Lee, N. H. (2010, 1 Jun 2010). *Mathematical modelling: An introduction.* Workshop session during the Mathematical Modelling Outreach, River Valley High School, Singapore.

Ng, K. E. D., & Lee, N. H. (2012). *Mathematical modelling: A collection of classroom tasks.* Singapore: Alston Publishing House Private Limited.

Niss, M. (2010). Modeling a crucial aspect of students' mathematical modeling. In R. Lesh, P. Galbraith, C. Haines & A. Hurford (Eds.), *Modeling students' mathematical modeling competencies: ICTMA 13* (pp. 43–59). New York: Springer.

Organisation for Economic Co-Operation and Development [OECD]. (2013). Draft PISA 2015 mathematics framework. (10 April 2013). Retrieved from http://www.oecd.org/pisa/pisaproducts/Draft%20PISA%202015%20Mathematics%20Framework%20.pdf.

Polya, G. (1957). *How to solve it*. Garden City, New York: Doubleday.

Stillman, G. A. (2010). Implementing applications and modelling in secondary school: Issues for teaching and learning. In B. Kaur & J. Dindyal (Eds.), *Mathematical applications and modelling: Yearbook 2010 of the Association of Mathematics Educators* (pp. 300–322). Singapore: World Scientific.

Stillman, G., Brown, J., & Galbraith, P. L. (2008). Research into the teaching and learning of applications and modelling in Australasia. In H. Forgasz, A. Barkatsas, A. Bishop, B. Clarke, S. Keast, W. T. Seah & P. Sullivan (Eds.), *Research in mathematics education in Australasia: New directions in mathematics and science education* (pp. 141–164). Rotterdam: Sense Publishers.

Tan, L. S., & Ang, K. C. (2013). Pre-service secondary school teachers' knowledge in mathematical modelling — A case study. In G. Stillman, G. Kaiser, W. Blum & J. Brown (Eds.), *Teaching mathematical modelling: Connecting to research and practice* (pp. 373–384). Dordrecht: Springer.

Section 1

Setting the Stage for Mathematical Modelling in Schools

CHAPTER 2

PROMOTION OF MATHEMATICAL MODELLING COMPETENCIES IN THE CONTEXT OF MODELLING PROJECTS

Gabriele KAISER

Susanne GRÜNEWALD

Keywords: Competencies, Projects, Germany, Realistic Modelling, Epistemological Modelling, Educational Modelling, Model Eliciting, Socio-Critical Modelling, Socio-Cultural Modelling, Cognitive Modelling.

2.1. Introduction

The relevance of mathematical modelling for mathematics education is currently consensus all over the world. Especially the promotion of modelling competencies, i.e., the competencies to solve real-world problems using mathematics, is accepted as central goal for mathematics education worldwide, especially if mathematics education aims to promote responsible citizenship. In many national curricula, such as the German or the US American national standards, modelling competencies play a decisive role pointing out that the importance of mathematical modelling is accepted at a broad international level. However, beyond this consensus on the relevance of modelling, it is still disputed, how to integrate mathematical modelling into the teaching-and-learning-processes, various approaches are discussed, either as short-term activities in one or two lessons during ordinary teaching time or as modelling projects, during which students tackle modelling problems over a longer period. More simple modelling activities seem to be possible to be dealt with in short-term activities whereas authentic, more complex modelling problems need intensive teaching time and therefore seem to be more suitable for more extensive modelling projects. These kinds of modelling projects have become popular in Germany in the last decade, promoted amongst others by activities at the Technical University of Kaiserslautern or the University of Hamburg.

In the following we will concentrate on these kinds of project activities and their approach to foster students' modelling competencies concentrated on the activities at the University of Hamburg, carried out amongst other by the authors. The first part of this chapter deals with the theoretical debate on mathematical modelling in school. Afterwards different kinds of modelling projects implemented at the University of Hamburg will be presented. Finally, selected results of the evaluations of these projects will be described and discussed.

2.2. Theoretical Debate on Mathematical Modelling in Mathematics Education

2.2.1. *Theoretical debate on mathematical modelling — historical development and current state*

The claim to teach mathematics in application-oriented way has been put forth already since the famous symposium "Why to teach mathematics so as to be useful" (Freudenthal, 1968, Pollak, 1968) has been carried out in 1968. Why and how to include applications and modelling in mathematics education has been the focus of many research studies since then. This high amount of studies has not led to a unique picture on the relevance of applications and modelling in mathematics education, in contrast the arguments developed since then remained quite diverse. In addition the discussion, how to teach mathematics so as to be useful did neither lead to a consistent argumentation. There have been several attempts to analyse the various theoretical approaches to teach mathematical modelling and applications and to clarify possible commonalities and differences. For example Kaiser-Meßmer (1986) distinguishes in her analysis from the beginning of the recent debate on modelling until the mid eighties of the last century two mainstreams within the international debate on applications and modelling, a so-called pragmatic perspective focusing on utilitarian or pragmatic goals with Pollak as protagonist and a scientific-humanistic perspective oriented more towards mathematics as a science and humanistic ideals of education with Freudenthal as main protagonist. The different goals have consequences for the way to include mathematical modelling, namely either based on cyclic model building processes as requested by Pollak (1969) or as complex mathematising interplay between mathematics and the real world as described by Freudenthal (1973).

In their extensive survey on the state-of-the-art Blum and Niss (1991) focus a few years later on the arguments and goals for the inclusion of applications and modelling and discriminate five layers of arguments such as the formative argument related to the promotion of general competencies, critical competence argument, utility argument, picture of mathematics argument and the promotion of mathematics learning argument. They make a strong plea for the promotion of three goals, namely, that student should be able to perform modelling processes, to acquire knowledge of existing models and to critically analyse given examples of modelling processes.

Based on this position they analyse the various approaches how to consider applications and modelling in mathematics instruction and distinguish six different types to including applications and modelling in mathematics instruction, e.g., the separation approach, separating mathematics and modelling in different courses or the two-compartment approach with a pure part and an applied part. A continuation of integrating applications and modelling into mathematics instruction is the islands approach, where small applied islands can be found within the pure course, the mixing approach is even stronger in fostering the integration of applications and modelling, i.e., newly developed mathematical concepts and methods are activated towards applications and modelling, whenever possible, however in contrast to the next approach, the mathematics used is more or less given from the outset. In the mathematics curriculum integrated approach, the problems come first and mathematics to deal with them is sought and developed subsequently. The most advanced approach, the interdisciplinary integrated approach, operates with a full integration between mathematics and extra-mathematical activities where mathematics is not organised as separate subject.

In their classification of the historical and more recent debate on mathematical modelling in school Kaiser and Sriraman (2006) point out, that there exist several perspectives on mathematical modelling in the international discussion on mathematics education. They proposed a framework for the description of the various approach, later modified by Kaiser *et al.*

(2007), which classifies these conceptions according to the aims pursued with mathematical modelling, their epistemological background and their relation to the initial perspectives.

The following perspectives were described, which continue positions already emphasised at the beginning of the modelling debate:

- Realistic or applied modelling fostering pragmatic-utilitarian goals and continuing traditions of the early pragmatically oriented approaches;
- Epistemological or theoretical modelling placing theory-oriented goals into the foreground and being in the tradition of the scientific-humanistic approach;
- Educational modelling emphasising pedagogical and subject-related goals, which are integrating aspects of the realistic/applied and the epistemological/theoretical approaches taking up aspects of a so-called integrated approach being developed at the beginning of the nineties of the last century (cf. Kaiser, 1995).

More recently the following approaches have been proposed:

- Model eliciting and contextual approaches, which emphasise problem-solving and psychological goals;
- Socio-critical and socio-cultural modelling fostering the goal of critical understanding of the surrounding world connected with the recognition of the cultural dependency of the modelling activities.

In addition as kind of a meta-perspective the following perspective is distinguished, which has been developed in the last decade reflecting demands on more detailed analysis of the students' modelling process and their cognitive and affective barriers.

- Cognitive modelling putting the analysis of students' modelling process in the foreground and the promotion of mathematical thinking processes.

2.2.2. *The modelling process as key feature of modelling activities*

As already mentioned the way, how mathematical modelling processes are understood, how the relation between mathematics and the "rest of the world" (Pollak, 1968), are described play a decisive role within the modelling debate. Modelling processes are differently used by the various perspectives and streams within the modelling debate, already since the beginning of the discussion. The perspectives described above developed different notions of the modelling process either emphasising the solution of the original problem, as it is done by the realistic or applied modelling perspective, or the development of mathematical theory as it is done by the epistemological or theoretical approach. So, corresponding to the different perspectives on mathematical modelling there exist various modelling cycles with specific emphasis, for example designed primarily for mathematical purposes, research activities or usage in classrooms (for an overview see Borromeo Ferri 2006).

Although at the beginning of the modelling debate a description of the modelling process as linear succession of the modelling activities were common as well or the differentiation between mathematics and the real world was seen more statically (e.g., by Burkhardt, 1981), nowadays, despite some discrepancies, one common and widespread understanding of modelling processes has been developed. In nearly all approaches the idealised process of mathematical modelling is described as a cyclic process to solve real problems by using mathematics, illustrated as a cycle comprising different steps or phases.

In the following one modelling cycle will be described in detail, which was developed by Blum (1995) and Kaiser (1996) within an integrated perspective and which is based amongst others on work by Pollak. This description contains the characteristics, which nowadays can be found in various modelling cycles.

The shown modelling cycle describes an idealised modelling process: the given real problem is simplified in order to build a real model of the situation, amongst other many assumptions have to be made, central influencing factors have to be detected. To create a mathematical model the real model has to be translated into mathematics. However, the distinction between a real and a mathematical model is not always well-defined, because the process of developing a real model and a mathematical model is interwoven, amongst others because the developed real problem is related to the mathematical knowledge of the modeller. Inside the mathematical model mathematical results are worked out by using mathematics. After interpreting the mathematical results the real results have to be validated as well as

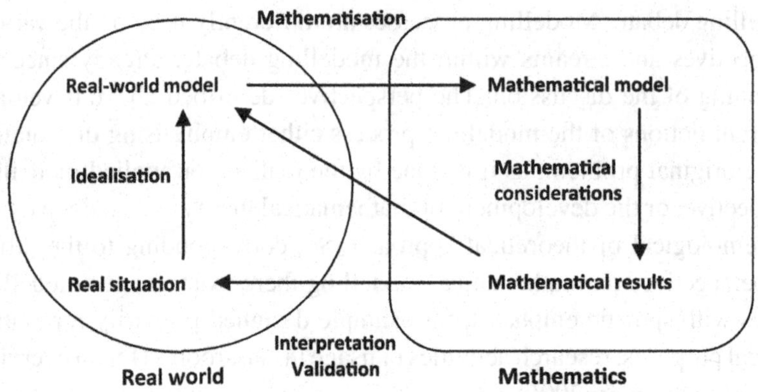

Figure 2.1. Modelling process from Kaiser (1995, p. 68) and Blum (1996, p. 18).

the whole modelling process itself. There may be single parts or the whole process to go through again.

The shown cycle idealises the modelling process. In reality, several mini-modelling-cycles occur, that are either worked out in linear sequential steps like the cycle or in a less ordered way. Most modelling processes include frequent switching between the different steps of the modelling cycles (Borromeo Ferri, 2011).

Other descriptions of the modelling cycle coming from applied mathematics, such as the one by Haines, Crouch, Davis (2000), emphasise the necessity to report the results of the process and include more explicitly the refinement of the model.

Perspectives putting cognitive analyses in the foreground, which have been described in the paragraph before, include an additional stage within the modelling process, the understanding of the situation by the students. The students develop a situation model, which is then translated into the real model, Blum in more recent work (e.g. 2011) or Borromeo Ferri (2011) have described modelling activities in such a way.

2.2.3. *Modelling competencies and their promotion*

A central goal of mathematical modelling is the promotion of modelling competencies, i.e., the ability and the willingness to work out problems with

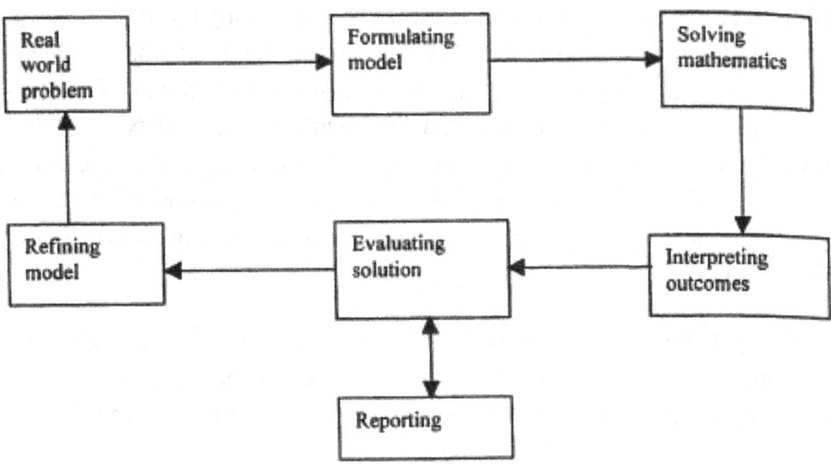

Figure 2.2. Modelling process from Haines, Crouch, Davis (2000, p. 3).

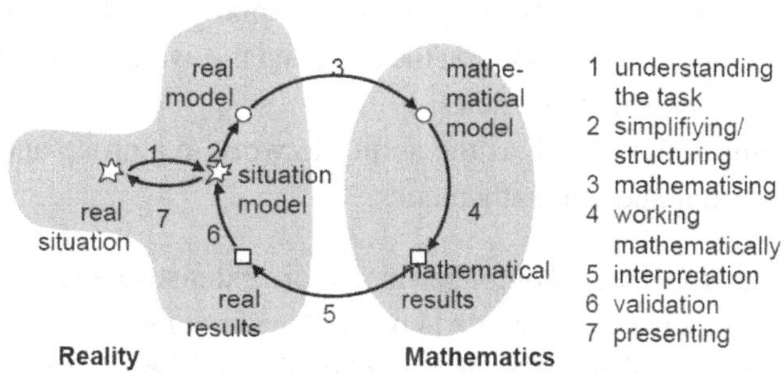

Figure 2.3. Modelling process by Blum (2011, p. 18).

mathematical means taken from the real world problem through mathematical modelling (cf. Maaß, 2004, 2006 for a detailed definition of modelling competencies). The definition of modelling competencies corresponds with the different perspectives of mathematical modelling and is influenced by the taken perspective (Zöttl, 2010). A distinction is made between global modelling competencies and sub-competencies of mathematical modelling. Global modelling competencies refer to necessary abilities to perform the whole modelling process and to reflect on it. The sub-competencies of mathematical modelling refer to the specific modelling cycle, they include the different competencies that are essential for performing the single steps of the modelling cycle (Kaiser, 2007). Based on the extensive study by Maaß (2004, 2006), own studies (Kaiser, 2007, Kaiser & Schwarz 2007), extensive work by Haines, Crouch & Davis (2000), Houston & Neill (2003) and further studies, which are summarised by Blomhøj (2011) and referring to the various types of the modelling cycle as described in Chapter 1.2.2., the following sub-competencies can be distinguished (Kaiser 2007, p.111):

- "Competency to solve at least partly a real world problem through mathematical description (that is, model) developed by oneself;
- Competency to reflect about the modelling process by activating meta-knowledge about modelling processes;
- Insight into the connections between mathematics and reality;
- Insight into the perception of mathematics as process and not merely as product;
- Insight into the subjectivity of mathematical modelling, that is, the dependence of modelling processes on the aims and the available mathematical tools and students competencies;
- Social competencies such as the ability to work in a group and to communicate about and via mathematics."

This list is far from being complete since more extensive empirical studies are needed to receive well-founded knowledge about modelling competencies; the study described in the next chapter aims to contribute to this goal.

Obviously the sub-competencies are an essential part of the modelling competencies. Referring to Maaß (2006), Zöttl (2010) or Stillman (2011), meta-cognitive competencies play a significant role within the modelling process as well. Missing meta-cognitive competencies may lead to problems during the modelling process, for example at the transitions between the single steps of the modelling cycle or in situations, where cognitive barriers appear (cf. Stillman, 2011).

In the discussion on the teaching and learning of mathematical modelling two different approaches of fostering mathematical modelling competencies can be distinguished: the holistic and the atomistic approach. The holistic approach assumes that the development of modelling competencies should be fostered by performing complete processes of mathematical modelling, whereby the complexity and difficulty of the problems should be matched to the competencies of the learners (Haines, Crouch & Fitzharris, 2003). The atomistic approach, however, assumes that the implementation of complete modelling problems, especially at the beginning, would be too time-consuming and not sufficiently effective at fostering the individual modelling competencies. Propagated is therefore the separate treatment of individual phases of the modelling process, i.e., individual fostering of sub-competencies of mathematical modelling (Blomhøj & Jensen, 2003).

Obviously these two different approaches necessitate different ways of organising the inclusion modelling examples in schools: the atomistic approach seems to be more suitable for a "mixing approach", i.e., "in the teaching of mathematics, elements of applications and modelling are invoked to assist the introduction of mathematical concepts etc. Conversely, newly developed mathematical concepts, methods and results are activated towards applicational and modelling situations whenever possible" (Blum & Niss, 1991, p. 61). The holistic approach can either be realised in "separation approach", i.e., instead "of including modelling and applications work in the ordinary mathematics courses, such activities are cultivated in separate courses specially devoted to them" (Blum & Niss, 1991, p. 60). Of course variations of this approach like the "two-compartment approach" or the "islands approach" described by Blum & Niss (1991) seem to be possible as well.

2.3. Promoting Modelling Competencies

Based on arguments taken from the realistic or applied perspective on modelling we claim the necessity to tackle authentic comprehensive modelling problems, in order to allow the students to gain insight into the richness and variety of mathematical modelling examples and processes and their relevance for their current and future lives (cf. Kaiser & Schwarz, 2011). In the following, we describe modelling projects, which foster modelling competencies under a holistic perspective in a separation or island approach. In addition, these projects allow the students to develop global modelling competencies and sub-competencies as well as meta-cognitive modelling competencies.

2.3.1. *Promotion of modelling competencies within modelling projects at the University of Hamburg*

Already for a decade, the University of Hamburg offers various modelling projects aiming for students at lower and upper secondary level either taking place at the university or in participating schools. As already mentioned the promotion of modelling competencies of the students as well the advancement of an adequate view on mathematics are the central goals of these projects. These various modelling projects can be assigned to the holistic approach to foster modelling competencies, as both projects aim to carry out complete modelling processes using complex modelling problems.

Since 2009, these activities are organised as modelling weeks, which take place twice a year. The modelling weeks are designed for students at upper secondary level from schools in Hamburg and its surrounding.

During the modelling weeks, the students tackle in small groups one authentic complex modelling problem developed out of the practice of applying mathematicians in industry for example the planning of wind parks, the chlorination of swimming pools or the production of microchips. The groups shall work as independent as possible, they are supervised by tutors, who only intervene by cognitive barriers, missing mathematical means or when the students are in cul-de-sac (for details of this approach see for example Kaiser & Schwarz, 2011).

In addition since 2010 the Working group on Didactics of Mathematics organises modelling activities with younger students at lower secondary level. These activities take place once a year with the whole age cohort of secondary schools in Hamburg. The current modelling days are designed for students of Grade 9. Due to the younger age of the students these modelling activities last only three days and deal with less ambitious examples, for example automatic watering of gardens or the planning of a bus line. The organisation of the work follows the same pattern, i.e. independent work by the students in small groups, supervised by tutors studying to become mathematics teachers.

2.3.2. *Evaluation of the promotion of modelling competencies*

The evaluation of modelling competencies and their development has been in the focus of empirical studies already for several years. Haines, Crouch &

Davis (2001) developed a modelling test in order to evaluate the efficiency of university modelling courses, this test was extended and refined by Houston & Neill (2003) as well as by Izard, Haines, Crouch, Houston & Neill (2003). The test intended to measure the development of the described sub-competencies related to the phases of a modelling process using multiple-choice items. This test was used by Kaiser (2007) for students at upper secondary level within the modelling projects at the University of Hamburg. Frejd & Ärlebäck (2011) used the same test in order to investigate the modelling competencies of Swedish upper secondary students. Other approaches such as the projects DISUM (Blum, 2011) or KOMMA (Zöttl, 2010) or activities by Maaß (2004, 2006), developed own tests considering meta-cognitive competencies and mathematical skills as well.

In order to investigate the effects of these new modelling projects established more recently at the University of Hamburg, aiming at the promotion of modelling competencies for students at lower secondary level the test developed by Haines and others was adapted for the study and complemented by additional interviews. In the following we describe the design of the study, the test developed and selected results.

2.3.2.1. Design of the study and the test

The evaluation study was carried out in 2010 within the frame of a master thesis (Grünewald, 2010). The evaluation aimed to answer mainly the following questions: To what extent can students' modelling competencies be fostered within a modelling project? Is there a relationship between the development of the students' modelling competencies and their mathematical abilities? Are there gender differences in the development of the students' modelling competencies? Do specific sub-competencies develop differently?

To answer these research questions the adapted modelling test used a pre- and post-format as in the original study. Each test-item tested one of the following sub-competencies, which do not cover all sub-competencies distinguished in the literature:

- Simplifying the real-world problem;
- Clarifying the goal;
- Defining the problem;

- Assigning central variables and their relations;
- Formulating mathematical statements;
- Selecting a mathematical model;
- Interpreting the solution in a real-world context;
- Validating the appropriateness of the solution.

Due to time restrictions it was not possible to test the global modelling competency with additional items. This gap has been closed in ongoing, not published research. Following the design of the original test multiple-choice items were used, four possible answers were given per item, one was correct, one was partly correct and two answers were wrong. Corresponding to the partial credit system of the modelling test, the individual response options of the different items were rated differently with 2 points, 1 point or 0 point. All in all, the highest possible score for one student was 16 points.

The original questions were replaced by easier items more appropriate for this age group. Figure 2.4 shows exemplarily one item used referring to the phase of identifying central variables and Figure 2.5 shows the item referring to the validation phase.

At the beginning and at the end of the modelling days 136 students (63 girls and 73 boys) filled in this questionnaire; to answer the ten questions of the test the students had 30 minutes time. Both questionnaires were divided

<table>
<tr><td colspan="2">Question 6: Refuelling
The free-lance lawyer Mrs. Mahnke lives in Treplin, 15 km away from the Polish border. For refuelling she drives to Poland, where there is a petrol station right at the border. In Poland one litre of petrol costs 1.02 Euro. In Treplin one litre of petrol costs 1.33 Euro. Is the drive profitable for Mrs. Mahnke?
Which of the following groups of factors is most important one for solving the task?</td></tr>
<tr><td>(A)</td><td>The size of the tank; the average gas consumption of the car; the hourly wage lost because of the drive.</td></tr>
<tr><td>(B)</td><td>The size of the tank; the gas consumption on country roads; the time of the day when she refuels.</td></tr>
<tr><td>(C)</td><td>The average gas consumption of the car in cities; the proportions of the tank; the opening hours of the petrol station.</td></tr>
<tr><td>(D)</td><td>The number of the people in the car; the road conditions to Poland; the time of the day when she refuels.</td></tr>
</table>

Figure 2.4. Assigning central variables and their relations.

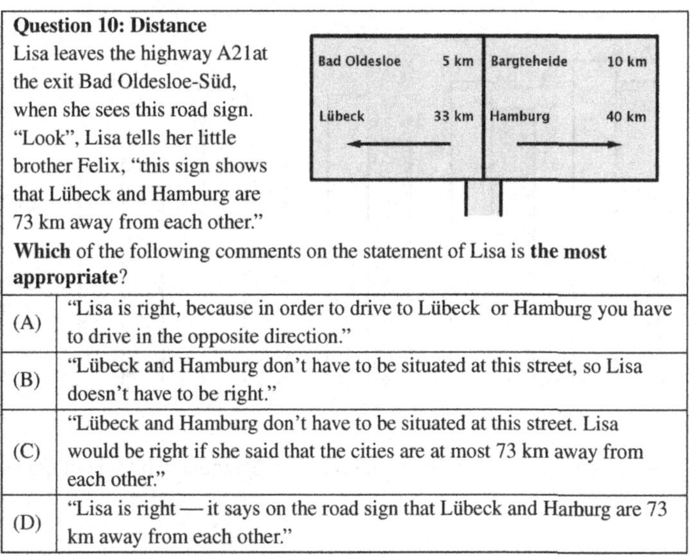

Question 10: Distance

Lisa leaves the highway A21at the exit Bad Oldesloe-Süd, when she sees this road sign. "Look", Lisa tells her little brother Felix, "this sign shows that Lübeck and Hamburg are 73 km away from each other."

Which of the following comments on the statement of Lisa is **the most appropriate**?

(A)	"Lisa is right, because in order to drive to Lübeck or Hamburg you have to drive in the opposite direction."
(B)	"Lübeck and Hamburg don't have to be situated at this street, so Lisa doesn't have to be right."
(C)	"Lübeck and Hamburg don't have to be situated at this street. Lisa would be right if she said that the cities are at most 73 km away from each other."
(D)	"Lisa is right — it says on the road sign that Lübeck and Harburg are 73 km away from each other."

Figure 2.5. Validating the appropriateness of the solution.

into two parts. Part A contained two questions about students' gender and previous grades in mathematics classes. Part B contained eight multiple-choice questions about modelling problems.

In addition to the modelling tests, five students (two girls and three boys with different mathematical skills) were interviewed separately at the beginning and at the end of the project referring to their given answers in the questionnaires; due to space limitations we will only in brief report on the results of these interviews.

2.3.2.2. Results concerning the promotion of modelling competencies

The analysis of the collected data took place on two levels: first the average overall scores of the students in the pre- and the post-test were compared with each other as well as potential relationships to gender or grade. Afterwards the average scores per item as well as the tendencies of the achieved overall scores were analysed.

In the pre-test the students reached an average overall score of 10.96 points and in the post-test 12.57 points. The girls achieved an average overall

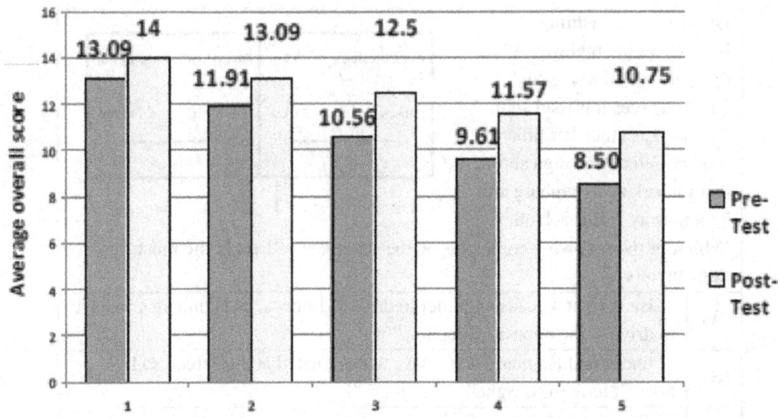

Figure 2.6. Comparison of the overall scores differentiated by school grades.

score of 10.98 points in the pre- and 12.7 points in the post-test. The boys gained an average overall score of 10.92 points in the pre- and 12.45 points in the post-test. Altogether, the average overall score of the students increased statistical significant from the pre- to the post-test, while the differences between the results of the girl and boys are not statistical significant.

The average overall scores of the students differentiated by their achievements in previous mathematics lessons are illustrated in Figure 2.6. The German school system differentiates achievements between 1 and 6 with 6 being the worst grade (not existent within this group of students). The figure shows that all students were able to achieve a higher average overall score in the post-test. Obviously, there is a correlation between the achieved score and the previous grade in mathematics, but the differences between the extents of improvement differentiated by grade are not statistical significant.

Significant improvements of the gained average scores per item could in addition be identified, except for item A9, which refers to the interpretation of the solution in a real-world context (see Figure 2.7). The highest increase of the achievements was found within items, which do test competencies that are usually not fostered by usual mathematical tasks, for example the validation competence. Significant differences between the results per item differentiated by gender or mathematical grade were not found.

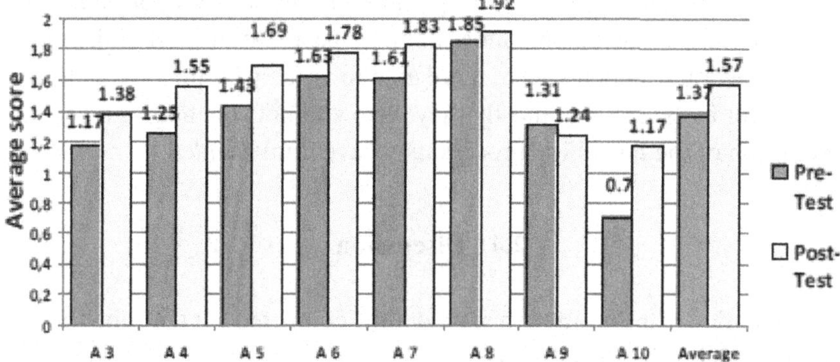

Figure 2.7. Comparison of the results per item.

Grade	Students with a higher score	Students with the same score	Students with a lower score
1	64%	27%	9%
2	59%	34%	7%
3	90%	8%	2%
4	82%	11%	7%
5	100%	0%	0%

Figure 2.8. Tendencies of the achieved overall scores according to grade.

Furthermore the data show the following tendencies of the achieved overall scores in the pre-test compared to the post-test: 77% of all students gained a higher overall score in the post- than in the pre-test, 18% achieved the same score and only 5% reached a lower score, with most closely corresponding results of the girls and the boys. In addition these tendencies can be differentiated by grade, which is illustrated in Figure 2.8. The table shows the percentage of students differentiated by grade who reached a higher/the same/a lower score in the post-test compared to the pre-test. The table shows that out of the group of students with lower grades a bigger part achieved a higher overall score than out of the group of students with higher grades.

The evaluation of the interviews supports the results of the evaluation of the modelling test. The responses of the students support the results concerning the improvement of the sub-competencies of mathematical modelling in the modelling tests, in particular of the validation competence.

During the first interviews at the beginning of the modelling days the students' answers are given on a more intuitive basis. At the end of the modelling project the students were more able to argue why they have chosen one answer in the test, but mostly they were still not able to give a general description of the modelling process they have implemented.

2.4. Discussion

The evaluation of the promotion of modelling competencies described above allows the following conclusions:

The modelling projects currently carried out at the University of Hamburg seem to be adequate to promote modelling competences even for younger students. The evaluation of the questionnaire shows a significant enhancement of the sub-competencies of mathematical modelling of the students'. Significant differences between the improvements differentiated by gender or mathematical abilities were not found.

Frejd & Ärlebäck (2011), who utilised the same test for a survey of the modelling competencies of Swedish upper secondary students, also did not make out any significant effect on the achieved average scores of the students with respect to gender. The correlation between the achieved score and the previous grade in mathematics corresponds to the results of Frejd & Ärlebäck (2011); in their study the students' grades had a significant effect on the students' overall scores too.

The results of this evaluation get in line with results of large-scale projects such as DISUM (Blum, 2011) or a study within the project KOMMA (Zöttl, 2010). These projects also aimed at fostering students' modelling competencies, but by integrating modelling tasks that can be treated within one lesson respectively by implementing heuristic worked-out examples. These project as well as Maaß (2004) stated that the integration of modelling problems into mathematics lessons leads to an increase of modelling competencies.

However the results must be treated cautiously, because the described developments are short-term effects of one modelling project reached with smaller group of students. In any case, the results of the evaluations show promising tendencies concerning the chances to promote modelling competencies using complex modelling examples.

References

Blomhøj, M. (2011). Modelling competency: Teaching, learning and assessing competencies – overview. In G. Kaiser, W. Blum, R. Borromeo Ferri & G. Stillman (Eds.), *Trends in teaching and learning of mathematical modelling* (pp. 343–349). New York: Springer.

Blomhøj, M. & Jensen, T. Højgaard (2003). Developing mathematical modelling competence: conceptual clarification and educational planning. *Teaching Mathematics and its Applications*, 22(3), 123–139.

Blum, W. (1996). Anwendungsbezüge im Mathematikunterricht – Trends und Perspektiven. In G. Kadunz, H. Kautschitsch, G. Ossimitz & E. Schneider (Eds.), *Trends und Perspektiven* (pp. 15–38). Wien: Hölder-Pichler-Tempsky.

Blum, W. & Niss, M. (1991). Applied mathematical problem solving, modelling, applications, and links to other subjects – state, trends and issues in mathematics instruction. *Educational Studies in Mathematics*, 22, 37–68.

Blum, W. (2011). Can modelling be taught and learnt? Some answers from empirical research. In G. Kaiser, W. Blum, R. Borromeo Ferri & G. Stillman (Eds.), *Trends in teaching and learning of mathematical modelling* (pp. 15–30). New York: Springer.

Borromeo Ferri, R. (2006). Theoretical and empirical differentiations of phases in the modelling process. *Zentralblatt fuer Didaktik der Mathematik*, 38(2), 86–95.

Borromeo Ferri, R. (2011). *Wege zur Innenwelt des mathematischen Modellierens*. Wiesbaden: Vieweg-Teubner.

Burkhardt, H. (1981). *The real world and mathematics*. Glasgow: Blackie.

Frejd, P. & Ärlebäck, J. B. (2011). First results from a study investigating Swedish upper secondary students' mathematical modelling competencies. In G. Kaiser, W. Blum, R. Borromeo Ferri & G. Stillman (Eds.), *Trends in teaching and learning of mathematical modelling* (pp. 407–416). New York: Springer.

Freudenthal, H. (1968). Why to teach mathematics so as to be useful. *Educational Studies in Mathematics*, 1(1/2), 3–8.

Freudenthal, H. (1973). *Mathematics as an Educational Task*. Dordrecht: Reidel.

Grünewald, S. (2010). *Eine empirische Studie zur Entwicklung von mathematischen Modelllierungskompetenzen*. Hamburg: University of Hamburg, unpublished Master thesis.

Haines, C. & Crouch, R. (2001). Recognizing constructs within mathematical modelling. *Teaching Mathematics and its Applications*, 20(3), 129–138.

Haines, C., Crouch, R. & Davis, J. (2000). *Mathematical modelling skills: A Research Instrument* (Technical Report No. 55). University of Hertfordshire: Dept. of Mathematics.

Haines, C., Crouch, R. & Fitzharris, A. (2003). Deconstruction mathematical modelling: Approaches to problem solving. In Q.-X. Ye, W. Blum, K. Houston & Q.-Y. Jiang (Eds.), *Modelling and mathematics education ICTMA 9. Applications in Science and Technology* (pp. 366–380). *Chichester: Horwood.*

Houston, K. & Neill, N. (2003). Assessing modelling skills. In S. J. Lamon, W. A. Parker & S. K. Houston (Eds.), *Mathematical Modelling: A way of life ICTMA11* (pp. 155–164). Chichester: Horwood.

Izard, J., Haines, C. R., Crouch, R. M., Houston, S. K & Neill, N. (2003). Assessing the impact of teaching mathematical modelling: some implications. In S. J. Lamon, W. A. Parker & S. K. Houston (Eds.), *Mathematical modelling: a way of life ICTMA11* (pp. 165–177). Chichester: Horwood.

Kaiser-Meßmer, G. (1986). *Anwendungen im Mathematikunterricht. Vol. 1 — Theoretische Konzeptionen*. Bad Salzdetfurth: Franzbecker.

Kaiser, G. (1995). Realitätsbezüge im Mathematikunterricht — Ein Überblick über die aktuelle und historische Diskussion. In G. Graumann *et al.* (Eds.), *Materialien für einen realitätsbezogenen Mathematikunterricht* (pp. 66–84). Bad Salzdetfurth: Franzbecker.

Kaiser, G. (2007). Modelling and modelling competencies in school. In C. Haines, P. Galbraith, W. Blum & S. Khan (Eds.), *Mathematical Modelling (ICTMA 12): Education, Engineering and Economics* (pp. 110–119). Chichester: Horwood.

Kaiser, G. & Sriraman, B. (2006). Theory Usage and Theoretical Trends in Europe — A Survey and Preliminary Analysis of CERME4 Research Reports. *Zentralblatt für Didaktik der Mathematik*, 38(1), 22–51.

Kaiser, G., Sriraman, B., Blomhøj, M. & Garcia, F. J. (2007). Report from the Working Group Modelling and Applications — Differentiating Perspectives and Delineating Commonalities. In D. Pitta-Pantazi & G. Philippou (Eds.), *Proceedings of the Fifth Congress of the European Society for Research in Mathematics Education* (pp. 2035–2041). Larnaca: University of Cyprus.

Maaß, K. (2004). *Mathematisches Modellieren im Unterricht. Ergebnisse einer empirischen Studie*. Hildesheim: Franzbecker.

Maaß, K. (2006). What are modelling competencies? *Zentralblatt für Didaktik der Mathematik*, 38(2), 113–142.

Pollak, H. O. (1968). On some of the problems of teaching applications of mathematics. *Educational Studies in Mathematics*, 1(1/2), 24–30.

Pollak, H. O. (1969). How can we teach applications of mathematics? *Educational Studies in Mathematics*, 2, 393–404.

Stillman, G. (2011). Applying Metacognitive Knowledge and Strategies in Applications and Modelling Tasks at Secondary School. In G. Kaiser, W. Blum, R. Borromeo Ferri & G. Stillman (Eds.), *Trends in Teaching an Learning of Mathematical Modelling: ICTMA14* (pp.165–180). Dordrecht: Springer.

Zöttl, L. (2010). *Modellierungskompetenz fördern mit heuristischen Lösungsbeispielen*. Hildesheim: Franzbecker.

CHAPTER 3

PROBLEM FINDING AND PROBLEM POSING FOR MATHEMATICAL MODELLING

Gloria STILLMAN

Problem finding and posing is an essential part of mathematical modelling and thus a necessary ingredient in any educational programme involving the teaching of modelling. When modelling, it is essential that modellers themselves are given the freedom to pose their own problems as they make the transition from the real world messy situation to a problem statement to begin to mathematise. This chapter provides teachers with ideas about where they might begin in teaching problem posing in the classroom where mathematical modelling will be a genuine focus.

Keywords: Modelling Process, Problem finding, Problem posing, Task Design.

3.1. Introduction

"Successful teaching and learning of mathematical modelling is ... about developing a different worldview as both a teacher and a student" (Stillman, 2010, p. 301). According to Mason (1984), in mathematics in classrooms "we wish to stimulate enquiry, because that can be followed by investigation and modelling" (p. 220). Thus, "a book full of questions ... run[s] entirely opposite to a classroom atmosphere of enquiry, of asking questions, since much of the force of a question lies in the asking" (p. 220). Humans are inveterate problem finders from an early age and the resolution of problems brings pleasure and fosters well being (Getzels, 1979). Once found, a problem needs to be posed in a productive manner so that it can be solved usefully. Problem posing activities underpin the development of a mathematisation disposition (Bonotto, 2010a) — a critical element in becoming a modeller (Bonotto, 2010b). Furthermore:

Problem posing is an important aspect of both pure and applied mathematics and an integral part of modelling cycles which require the mathematical idealisation of the real world phenomenon (Christou *et al.* 2005). For this reason, problem posing is of central importance in the discipline of mathematics and in the nature of mathematical thinking and it is an important companion to problem solving (Bonotto, 2010a, pp. 403–404).

3.2. What is Problem Posing?

Problem posing in a real world situation occurs when a problem is formulated in such a way that it is amenable to mathematical analysis. So there is a problem and only afterwards it is formulated, that is, posed, in a certain way. There are many dilemmas in the world around us but these are often amorphous situations which have to be "transformed into a situation where a problem for solution emerges" (Getzels, 1979, p. 168). This process resulting in a problem to be solved is referred to as "problem finding" (Dillon, 1982; Getzels, 1979; Getzels & Czikszentmihalyi, 1976) although at times the terms problem posing and the much less used, problem finding, are blurred. Dilemmas are part of messy real world situations. Unfortunately, these dilemmas usually do not cooperatively present themselves as problems in a form that we are able to resolve or even think about in a sensible fashion to model. Finding or generating the problem is a crucial cognitive step (Getzels, 1979, p. 168). Once a problem has been found it needs to be posed. The posing (i.e., specification or formulation) of the problem can occur in many different ways resulting in many different problems posed. "Question asking" is often associated with problem finding and posing but not all problems come with questions just as not all questions refer to problems; but to begin finding a problem and posing it, questions are a useful pedagogical tool. To start problem posing in the classroom, it is as simple as using a digital photograph such as this dinosaur fountain shown in Figure 3.1 and asking: What mathematical problem could you pose?

Similarly, the Emu boot in Figure 3.2 could be used as a stimulus for problem posing. Problems that come to mind include: *How tall would the person be who could wear such a boot? Considering the shortness of the*

Figure 3.1. Dinosaur fountain in Santa Monica, California.

Figure 3.2. The Emu boot in Ballarat, Australia.

Given the numbers 47, 84, 83, 95, 48 and 36, choose 2 of these and add them. The answer is one of the other numbers. Can the same be done for subtraction?

Figure 3.3. Example problem solving task.

foot of the boot, would such a person be able to be supported by the small feet that could fit in a pair of these boots?

3.3. Modelling and Problem Solving

Within the scheme of things it is natural to ask how modelling differs from problem solving. Problem solving is the application of mathematical reasoning to the development of a solution or solutions to a given problem. Problem solving can be purely mathematical and be quite abstract as shown in Figure 3.3. Mathematical modelling, on the other hand, involves using mathematical concepts, structures and relationships to describe and characterise, or model, a real-world situation in a way that it captures its essential features. Problem solving techniques will be a part of that process so the two are closely related.

In modelling it is essential that the modellers pose their own problems and this can easily be done by modellers of all ages as shown in this poster (Figure 3.4) from the 2008 A.B. Paterson College Gold Coast Modelling Challenge. Here the group of modellers has posed the problem in the form of a question: *Which is the best design for the blades of a wind generator?* The question posed by school students is both a genuine real-world task and of research interest to scientists is shown by the experimental wind turbines at Riso Research Station in Denmark where I shared accommodation on my sabbatical in 2007 with research students investigating this exact same problem albeit with three dimensional designs. Thus, in accordance with Niss' definition (1992), this is an authentic problem as it is what research scientists dealing with alternative energy sources might meet in their daily work.

3.4. Modelling as a Process

Many diagrams have been used over the years to capture the essence of what happens during mathematical modelling to act as a scaffold in talking about

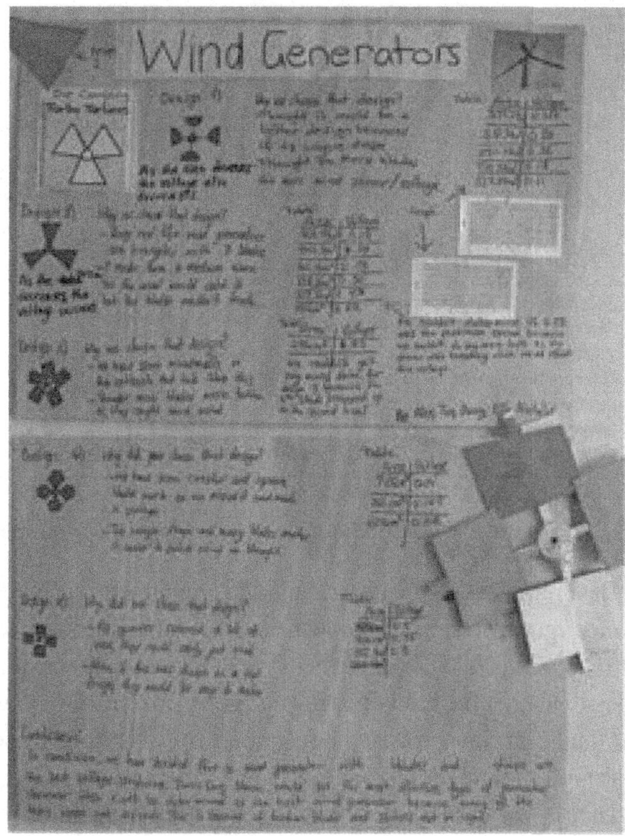

Figure 3.4. Poster from A.B. Paterson Mathematical Modelling Challenge.

Figure 3.5. Wind turbines at Riso Research Station, Denmark.

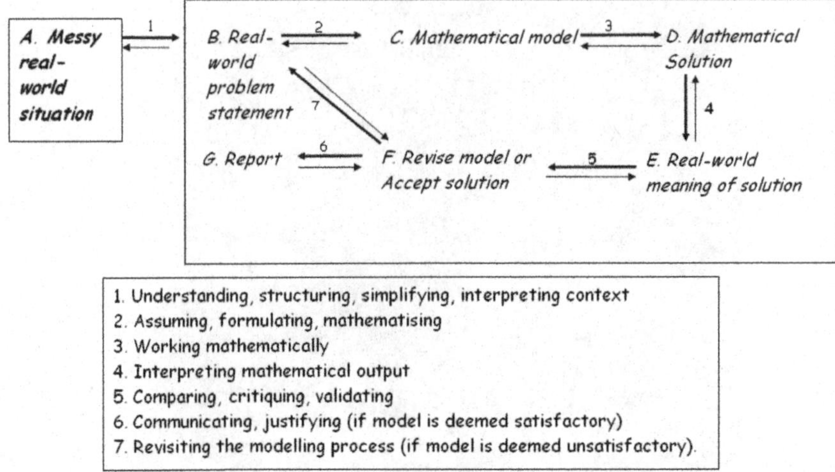

Figure 3.6. Modelling cycle from Stillman, Galbraith, Brown and Edwards (2007).

modelling, designing tasks for modelling and implementing modelling in the classroom. Figure 3.6 is a version of the cycle used in Stillman, Galbraith, Brown and Edwards (2007). The entries A–G represent stages in the modelling process, where the thicker arrows signify transitions between the stages. The total solution process is described by following these arrows clockwise around the diagram from the top left. It culminates either in the report of a successful modelling outcome, or a further cycle of modelling if evaluation indicates that the solution is unsatisfactory in some way. The kinds of mental activity that individuals engage in as modellers attempt to make the transition from one modelling stage to the next are given by the broad descriptors of cognitive activity 1 to 7 in Figure 3.6. The light reverse arrows emphasise that the diagram is a simplification of the modelling process which is far from linear, or unidirectional. The light arrows indicate the presence of reflective metacognitive activity (see Stillman, 2010, for further details). Of particular interest in problem finding and problem posing is transition 1 where the product is a real-world problem statement arising from the real-world situation and all its complexity and transition 2 which begins with specifying assumptions and formulating.

Modelling involves many subskills or actions which have been identified by various researchers over the years. The ability of an individual to use these subskills in a modelling situation is referred to as "mathematical modelling competency" (Niss, Blum, & Galbraith, 2007, p. 12). These subskills[1] include: (a) formulating a specific question to be answered mathematically, (b) specifying assumptions, (c) identifying the important information or variables, (d) modelling different aspects of relationships, (e) generating relationships, (f) recognising patterns and relationships, (g) selecting relationships, (h) making estimates, (i) validating results, (j) interpreting results and (k) communicating results. At every stage along the way, modellers need to reflect on what they are doing. It is important to continually link the mathematics back to the real situation, and vice versa.

Mathematical modelling explicitly uses real-world contexts that elicit the creation of useful models and draw upon several topic areas not only from mathematics but also from other disciplines. Thus modelling promotes interdisciplinarity (English, 2009; Galbraith, 1998). Modelling problems encourage the development of generalisable models that transcend the specifics of one situation. Reflection on the models built is thus crucial in the modelling process. This is not to say that all models are generalisable, nor should they be. There are many authentic modelling examples in biology, environmental science, geography and economics, for example, where the focus is capturing the specifics of the situation. Later in this chapter a height predictor model based on length of the right foot and age will be presented. The model clearly cannot transcend the situation for which it is constructed from humans to horses, for example. This needs to be borne in mind so that as teachers we do not fall into the temptation of following didactical aims at the expense of the authenticity of the situation.

3.5. Problem Finding and Posing Preparing for Mathematising

Transitions 1 and 2 of the modelling cycle are notoriously difficult (see, e.g., Galbraith & Stillman, 2001; Stillman & Galbarith, 2003) and have been the subject of theoretical examination for quite some time (e.g., Lambert *et al.*, 1989; Treilibs, Burkhardt, & Low, 1980). A recent contribution by

[1] Others such as Maaß (2006) have grouped subsets of these subskills into different categories which they refer to sometimes as competencies when dealing with modelling holistically and at other times as sub-competencies when considering what is needed to make a single transition in the modelling cycle.

Niss (2010) regarding mathematisation processes opens up possibilities for useful discussion about reasons for this difficulty and considerations for teaching.

Niss introduces the idea of "implemented anticipation" (p. 56) which appears to this reader to be a subset of Treilibs' "sense of direction" discussed and exemplified in Maaß (2006). By "implemented anticipation" Niss means "the modeller has to be able to anticipate potentially difficult subsequent steps, and to implement this anticipation in terms of decisions and actions" framing subsequent steps in the modelling cycle (p. 56), that is, making the transitions in Figure 3.6. According to Niss' model of students' mental processes involved in mathematisation (2010, pp. 56–57), in structuring the real world situation in preparation for mathematising it, the processes of problem generation and problem posing, during which the elements of interest and questions arising are specified, must be informed by their anticipated usefulness to the particular modeller in mathematising. Mathematisation in turn involves anticipation of mathematical representations which will capture the situation in this way but this also involves anticipating the subsequent use these representations and the resulting model will be in the solving (i.e., in transition 3 in Figure 3.6).

To be able to develop this foresight, students would need to develop mathematical knowledge relevant to the situation and have nuanced experiences of working with various mathematical strategies and representations in a variety of realistic and full-blown modelling contexts. Clearly, experience builds experience. This can include experience in problem finding and problem posing activities as ends in themselves where the mathematisations are foreshadowed but not implemented. Classroom discussion with both peer and teacher input is useful here. Additionally, students need to be involved in full modelling tasks or projects where the problem finding and problem posing are an integral part, not focused on separately.

3.6. Designing Tasks: Where to Start?

In designing tasks to begin as teachers we all have to keep our eyes and ears open for suitable contexts which stimulate the posing of problems. Two such contexts are depicted in the photographs in Figure 3.7. Once a situation has captured your interest, then you need to say to yourself: What question

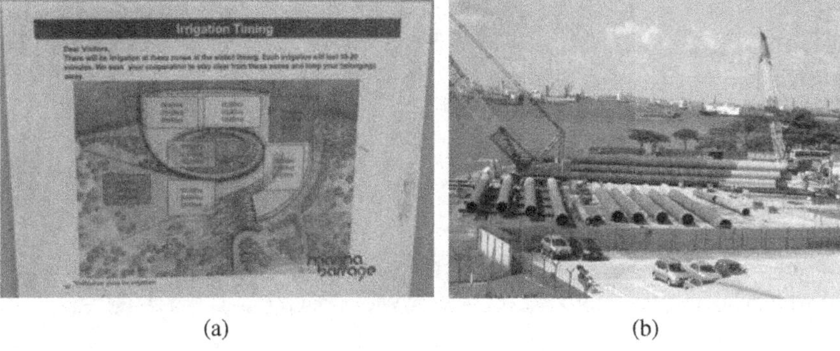

(a) (b)

Figure 3.7. Digital photographs as a source for beginning problem posing.

should I ask about this situation? We will then worry about whether it can be mathematised or is suitable for modelling. Being able to select between situations that lend themselves to mathematical analysis or not is a part of the skill toolkit that needs to be developed by students so let them pose their problems first then consider whether these can be mathematised. With experience, they will come to realise the importance of Niss' "implemented anticipation" (2010, p. 56) in the posing.

3.6.1. *Some examples*

A visit to Marina Barrage and the NEWater exhibit raises a range of possibilities. As shown in Figure 3.8 there is an impressive display of plastic water bottles that could be used as a stimulus for problem posing. Obvious questions are: *How high is the display? How many bottles did it take to make it? How much water would those bottles hold?* However, these are not burning questions that we all really need to know or even want to know the answers to. Contemplating further, the bottles are single-use which gives rise to the possibilities of considering the Mywaterbottle Programme as a source for further, but more authentic, problems to pose.

3.6.1.1. Posing

Mywaterbottle Programme is designed to reduce single-use bottled water. In 2007 Singaporeans consumed about 64 million litres of bottled water.

(a) (b)

Figure 3.8. Water bottle display at Marina Barrage.

They spent \$98 million buying single-use bottled water. It has been suggested that if 10% of Singaporeans started using refillable water bottles instead, this could save 4 million bottles of water per year. Thus, a more interesting problem that could be posed is: *What is the impact on our environment of single-use water bottles?* What sort of information might we need to investigate this mathematically in a way that involves mathematical modelling?

To begin with, the carbon footprint from production to disposal of plastic drink bottles is as follows:

- 302 g CO_2 eq. 0.5 L bottle
- 573 g CO_2 eq. 1 L bottle
- 801 g CO_2 eq. 1.5 L bottle.

Furthermore, in 2001 Singapore rolled out its National Recycling Programme to encourage residents to maximise recycling. Plastic, glass, paper, old clothing and metal waste began to be collected fortnightly. The recycling rate in Singapore for 2010, for example, was 58% with a further 40% incinerated at four waste-to-energy plants and 2% landfilled at Semakau Landfill. That was 6,517,000 tonnes of waste generated in Singapore in 2009. This amounted to 1280 kg per person.

What problems could we pose in order to answer our broader problem posed above: *Will Singapore be able to reach its Zerowaste Targets? Will the lifespan of Semakau landfill be able to be extended another*

Table 3.1. *Singapore Population 1960–2011.*

Year	Population	Year	Population
1960	1646400	2004	4166700
1970	2074500	2005	4265800
1980	2413900	2006	4401400
1990	3047100	2007	4588600
2000	4027900	2008	4839400
2001	4138000	2009	4987600
2002	44176000	22010	55076700
2003	44114800	22011	55183700

Source: Singapore Department of Statistics.

Table 3.2. *Waste Statistics 1998–2010.*

Year	Recycled Waste (t)	Disposed Waste (t)	Total Waste (t)	Recycling Rate (%)
1998	1860000	2840000	4700000	39.6
2000	1857300	2797300	4654600	40
2001				44
2002				45
2003	2220000	2510000	4730000	47
2004	2310000	2480000	4790000	48
2005	2540000	2470000	5010000	49
2006	2660000	2560000	5220000	51
2007	3030000	2570000	5600000	54
2008	3342600	2627600	5970200	56
2009	3485200	2628900	6114100	57
2010	3757500	2759500	6517000	58

Source: Singapore National Environment Agency.

50 years? To answer these questions some further data would be useful about two very important variables — trends in the population growth in Singapore (see Table 3.1) and the disposal and recycling of waste (see Table 3.2).

Thus, from a simple photographic stimulus we have developed a series of more in-depth situations to investigate and model through the problems we have posed. This process of drilling down from superficial surface questions to more in-depth questions develops modellers' insightfulness in problem posing.

3.6.1.2. Possible sources

An example of using a scientific article as a stimulus is estimation of height or weight by foot length using a statistical model. Such a model could be of use by the police in identifying victims of crime or accident or in cases of natural disaster where only fragments of a body might be found. Problems posed could be: *Are we able to predict the height of adolescents based on foot length data using either foot? Can the same model be used effectively with both males and females? Are the models specific to different groups of people? Can a good model be made with a diverse group of people? Can foot length be used to predict age? Can other foot dimensions such as breadth be used to predict height of a person?* Clearly, this is a rich context for problem posing. Grivas *et al.* (2008) developed several models from a sample of 5093 Greek school children aged 8 to 14 years. Their model having the greatest predictive value was described by the formula:

$$\begin{aligned}\text{height (cm)} &= 34.113 + 3.716 \times (\text{right foot length in cm}) \\ &\quad + 1.558\ (\text{if a girl}) + 2.499 \times (\text{age in years}).\end{aligned}$$

It was also found that the models which contained right rather than left foot length as an explanatory variable predicted height more accurately. Reading excerpts from this article or only the abstract would be sufficient as a stimulus for a problem posing activity.

A newspaper example relates to the often-used Big Foot Problem (Lesh & Harel, 2003, pp. 165–168) but with a paw print rather than a footprint or shoe impression. The text shown in Figure 3.9 about the discovery of "big paw prints" in rural Queensland was accompanied by a photograph of a man holding a plaster cast of a very large paw print which had been found on his rural property and a diagram showing the dimensions of a "normal" cat (75 cm long, 25 cm high), male panther (2 m long, 75 cm high) and a cow (2.6 m long, 1.52 m high). Big cats are not native to Australia. Again, the newspaper article is sufficient stimulus for students to pose problems, some of which will be able to be mathematised and require modelling.

Another example, the length of tape in a standard dental tape dispenser that can be commercially purchased in supermarkets and pharmacies can be used to estimate the amount of tape in a market sample small dispenser given to dentists (see Figure 3.10a for examples). Previous experience with

Big cat discovery casts little doubt

A GIANT cat is stalking wildlife in rural Glenwood, north of Gympie, and Colin Rossow reckons he has the proof — a plaster cast of a paw print bigger than his fist.

Mr Rossow, 67, who retired to his rural acreage in Pepper Rd 16 months ago, said there was no doubt some sort of panther size cat was living in the area.

"I grew up in the bush and I know when big cats are around," he said. "I've heard this cat before. It's not a kangaroo or a dog ... it's a caterwauling and sometimes it sort of grunts and chuffs."

Mr Rossow said his suspicions were confirmed when he was using his excavator to move stumps after heavy rain last week and spotted big paw prints in the mud.

"There was still this one big print where a kangaroo had come around the corner fast and sort of half slipped and the cat had come around after it and leaned into the corner with its front right paw," he said.

"There was one, big, perfect print and I had some gyprock plaster in the shed so I got a cast of it."

The cast shows a lion-sized paw print measuring 14.5 cm long by 13 cm wide with a deep pad and four toes with retractable cat-like claws.

Local legend also includes sightings of a black panther-like creature in the Gayndah area, northwest of Gympie.

Figure 3.9. Excerpts from newspaper article by Green (2009).

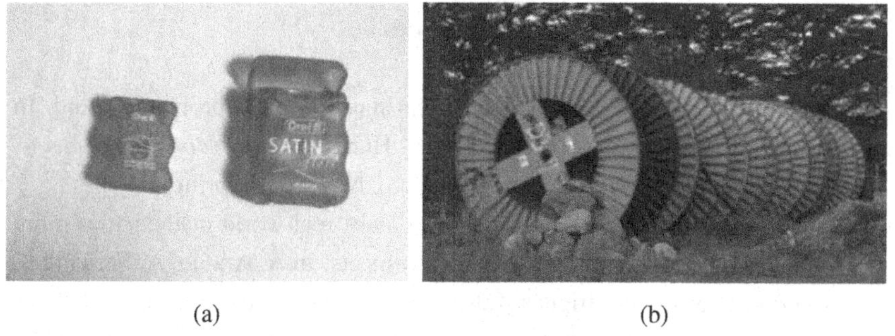

(a) (b)

Figure 3.10. (a) Dental tape dispensers. (b) Cable drums at Acciona Energy, Ballarat.

posing relevant mathematical questions in this activity can then be built on in later problem posing or full modelling activities when students are presented with photographs of large cable drums used by electricity authorities, telecommunications companies, city councils or contractors (see Föerster & Kaiser, 2010, for a modelling example using this context).

3.7. Concluding Remarks

In this chapter it has been argued that problem finding and problem posing are essential ingredients in any educational programme in schools promoting mathematical modelling. The fact that modellers find and pose their own problems to solve is one of the characteristic features of mathematical modelling. Several suggestions have been offered with regard to how teachers might begin to provide problem finding and problem posing activities in their classrooms as well as likely sources of suitable situations or objects to model. By providing students in school with many opportunities to find and pose problems for themselves, it is envisaged that they will begin to adopt a mathematisation disposition where "mathematics is indispensible as a way of knowing about the world in which they live and about the complex phenomena that affect the quality of their lives" (Lamon, Parker, & Houston, 2003, p. ix).

References

Bonotto, C. (2010a). Realistic mathematical modeling and problem posing. In R. Lesh, P. Galbraith, C. R. Haines & A. Hurford (Eds.), *Modeling students' mathematical competencies* (pp. 399–408). New York: Springer.

Bonnotto, C. (2010b). How is it possible to make real world mathematics more visible: Some results from two Italian projects. In A. Araújo, A. Fernandes, A. Azevedo & J. F. Rodrigues (Eds.), *Conference Proceedings of EIMI 2010: Educational Interfaces between Mathematics and Industry* (pp. 107–116). Lisbon: Centro Internacional de Matemática & Bedford, MA: COMAP.

Dillon, J. T. (1982). Problem finding and solving. *Journal of Creative Behaviour*, 16(2), 97–111.

English, L. (2009). Promoting interdisciplinarity through mathematical modeling. *ZDM — The International Journal on Mathematics Education*, 41(1/2), 161–181.

Föerster, F. & Kaiser, G. (2010). The cable drum — description of a challenging mathematical modeling example and a few experiences. In B. Kaur & J. Dindyal (Eds.), *Mathematical Applications and Modeling: Yearbook 2010 Association of Mathematics Teachers* (pp. 276–299). Singapore: World Scientific.

Galbraith, P. (1998). Cross-curriculum applications of mathematics. *International Review on Mathematical Education*, 30(4), 107–109.

Galbraith, P. & Stillman, G. (2001). Assumptions and context: Pursuing their role in modeling activity. In J. F. Matos, W. Blum, S. K. Houston & S. P. Carreira (Eds.), *Modelling and Mathematics Education* (pp. 300–310). Chichester, UK: Horwood.

Getzels, J. W. (1979). Problem finding: A Theoretical note. *Cognitive Science*, 3(2), 167–179.

Getzels, J. W. & Csikszentmihalyi, M. (1976). *The Creative Vision: A Longitudinal Study of Problem Finding in Art.* New York: John Wiley.

Green, G. (2009, January 14). Big cat discovery casts little doubt. *The Courier Mail*, p. 17.

Grivas, T. B., Mihas, C., Arapaki, A. & Vasiliadis, E. (2008). Correlation of foot length with height and weight in school age children. *Journal of Forensic and Legal Medicine*, 15(2), 89–95.

Lambert, P., Steward, A. P., Manklelow, K. I. & Robson, E. H. (1989). A cognitive psychology approach to model formulation in mathematical modeling. In W. Blum, J. S. Berry. R. Biehler, I. D. Huntley, G. Kaiser-Messmer & L. Profke (Eds.), *Applications and Modelling in Learning and Teaching Mathematics* (pp. 92–97). Chichester, UK: Horwood.

Lamon, S., Parker, B. & Houston, K. (2003). Mathematical modeling: A way of life. In S. Lamon, B., Parker & S.K. Houston (Eds.), *Mathematical Modeling: A Way of Life ICTMA 11* (pp. ix-x). Chichester, UK: Horwood.

Lesh, R. & Harel, G. (2003). Problem solving, modeling, and local conceptual development. *Mathematical Thinking and Learning*, 5(2/3), 157–189.

Maaß, K. (2006). What are modeling competencies? *ZDM*, 38(2), 113–142.

Mason, J. H. (1984). Modelling: What do we really want students to learn? In J. S. Berry, D. H. Burghes, I. D. Huntley, D. J. G. James & A. O. Moscardini (Eds.),

Teaching and Applying Mathematical Modelling (pp. 215–234). Chichester, UK: Horwood.

Niss, M. (1992). *Applications and Modelling in School Mathematics: Directions for Future Development.* Roskilde, Denmark: IMFUFA, Roskilde University Centre.

Niss, M. (2010). Modeling a crucial aspect of students' mathematical modeling. In *Modeling Students' Mathematical Modeling Competencies: ICTMA13* (pp. 43–59). New York: Springer.

Niss, M., Blum, W. & Galbraith, P. (2007). Introduction. In W. Blum, P. Galbraith, H-W. Henn & M. Niss (Eds.), *Modelling and Applications in Mathematics Education: The 14th ICMI Study* (pp. 3–32). New York: Springer.

Stillman, G. & Galbraith, P. (2003). Towards constructing a measure of the complexity of applications tasks. In S. Lamon, B., Parker & S. K. Houston (Eds.), *Mathematical Modeling: A Way of Life ICTMA 11* (pp. 179–188). Chichester, UK: Horwood.

Stillman, G. (2010). Implementing applications and modeling in secondary school: Issues for teaching and learning. In B. Kaur & J. Dindyal (Eds.), *Mathematical Applications and Modelling: Yearbook 2010 Association of Mathematics Educators* (pp. 300–322). Singapore: World Scientific.

Stillman, G., Galbraith, P., Brown, J. & Edwards, I. (2007). A framework for success in implementing mathematical modelling in the secondary classroom. In J. Watson & K. Beswick (Eds.), *Mathematics: Essential Research, Essential Practice.* Proceedings of the 30th annual conference of the Mathematics Research Group of Australasia (MERGA) (Vol. 2, pp. 688–707). Adelaide: MERGA.

Treilibs, V., Burkhardt, H. & Low, B. (1980). *Formulation Processes in Mathematical Modelling*. Nottingham, UK: Shell Centre for Mathematical Education.

CHAPTER 4

MATHEMATICAL MODELLING IN SINGAPORE SCHOOLS: A FRAMEWORK FOR INSTRUCTION

ANG Keng Cheng

In Singapore, unlike in many other parts of the world, the practice of mathematical modelling in the classroom is still in its infancy stage. However, interest in this area has grown rapidly in the last few years, and many local mathematics teachers and teacher educators have been exposed to a variety of views, interpretations and philosophies of modelling in schools. In this chapter, a framework for the practice of mathematical modelling in Singapore schools is proposed along with examples on how the framework may be used as a guide to design modelling learning experiences in local schools.

Keywords: Modelling Framework, Secondary Schools, Teacher Education.

4.1. Introduction

Mathematical modelling has a relatively short history in the Singapore mathematics education journey. In fact, Ang (2001) was probably the first to propose introducing mathematical modelling in Singapore schools. He argued that since mathematical problem solving is a central theme in Singapore's school mathematics curriculum, it is natural to expect part of the delivery to focus on applying mathematics in practical and real-life situations.

Other suggestions and examples on how mathematical modelling might be made relevant and possible in Singapore classrooms soon followed. These include applying the logistic equation to model the 2003 outbreak of the Severe Acute Respiratory Syndrome, or SARS, in Singapore (Ang, 2004), and studying how equations can be used to construct a model for traffic flow (Ang & Neo, 2005). Suggestions on how technology can support teaching of mathematical modelling in the classroom were also made (Ang, 2006; Ang, 2007; Ang, 2010a), and resources for teachers had come in the form of booklets containing collections of modelling and application activities (Ang, 2009; Dindyal 2009).

In the last two years, conscious efforts have been taken to promote the teaching of mathematical modelling among teachers. These include a Mathematics Teachers' Conference with a focus on mathematical applications and modelling in 2009 and a Mathematical Modelling Outreach programme in 2010, both of which were organised by the Mathematics and Mathematics Education (MME) group of the National Institute of Education (NIE), Singapore. In addition, a few local schools have also taken the initiative to run modelling workshops and challenges at cluster or zone levels.[1] In addition, an in-service programme on mathematical modelling for the secondary classroom was mounted recently to help build the capacity of practising teachers in carrying out modelling activities.

In general, it is therefore quite clear that mathematics teachers in Singapore are willing and keen to try and engage students in mathematical modelling activities in the classroom. However, what they are uncertain about is how to design and carry out modelling lessons in an organised and effective manner. In this chapter, we propose a framework that may be used to help teachers to systematically devise and develop learning experiences in mathematical modelling suitable for the classroom.

In the sections to follow, unless otherwise stated, all references to the mathematics teacher, the mathematics curriculum and classroom practices are made in the context of the situation in Singapore.

4.2. What Teachers Really Need

While workshops and special events on mathematical modelling may raise awareness among local teachers, they may not be as useful in helping teachers take the first step in actually preparing and implementing a modelling lesson or task. It is fair to say that a typical mathematics teacher in Singapore faces many challenges in the area of teaching mathematical modelling. These challenges include a lack of ready and relevant resources, a set of

[1] In Singapore, schools are grouped together in clusters and zones, according to their geographic locations.

good exemplars of modelling problems, and the resistance from students to engage in activities not directly related to examinations and assessment (Ang, 2010b).

Despite these challenges, it is encouraging to note that many teachers are keen and prepared to make an attempt to include mathematical modelling in their teaching curriculum. This can be seen in the very positive response to the events and in-service courses related to the teaching of mathematical modelling mounted in the last few years. Perhaps the one important missing piece that local mathematics teachers badly and urgently need is a practical framework that can serve as a guide in preparing mathematical modelling lessons, activities or learning experiences in the classroom.

It should be noted that a similar and related framework has been proposed. Stillman *et al.* (2007) outlined a structural framework to support the implementation of mathematical modelling in Australian schools. This framework is based largely on the widely accepted modelling cycle; in essence, it serves as a guide for the teacher to identify possible student blockages moving from one part of the modelling cycle to the next. While the framework may have been successfully applied by Stillman and her colleagues as reported, it appears to be meant more for the researcher, curriculum designer or an experienced teacher to identify particular modelling competencies needed to complete a modelling task. In that sense, such a sophisticated tool may not be suitable for the Singapore mathematics teacher, who is, in more ways than one, usually a novice or beginner in the teaching of mathematical modelling.

In the same way, while the set of design principles proposed and advocated by Galbraith (2006) may be useful for generating real life modelling problems and tasks, it may prove too advanced a tool for a beginner teacher modeller. Perhaps before one could use and apply these principles of design, one should be guided by a simple framework that focuses on turning an idea into a modelling lesson. These design principles can later be applied when one chooses to design a more complete modelling task.

A framework will also help teachers develop a shared understanding of what it means by teaching mathematical modelling in the classroom. Currently, among the teachers, it seems there is a plethora of interpretations of what constitutes mathematical modelling. This can be confusing and it is time we establish an understanding that is both useful and practical for the local schools. A framework will also provide teachers with a common language with which to discuss ideas and deliberate on issues related to mathematical modelling. This can lead to a more structured and professional interaction that may in turn lead to improving the current state of affairs.

4.3. Three Levels of Learning Experiences

Before discussing the proposed framework for instruction, it is perhaps important to define the scope within which one can properly organise or plan a mathematical modelling lesson in our classroom. A novice teacher in mathematical modelling would probably require a more structured approach to planning a successful learning experience in the classroom.

For this purpose, it is proposed that classroom learning experiences in mathematical modelling be classified into three levels, aligned to particular levels of cognitive demands and expectations.

Level 1

At the most basic level (Level 1), the focus is on acquiring skills that may be directly or indirectly related to mathematical modelling. These could be purely mathematical skills presented in a modelling context, or they may be specific skills that are often used in mathematical modelling activities. For example, a lesson on a specific function and its graph applied to a real life problem (or set in a real-life context) could be considered a Level 1 modelling learning experience.

Another Level 1 modelling activity could be a lesson on curve fitting using a computing or IT tool. In real-life problems, there is often the need to collect data and then to find suitable functions with suitable parameter values that best fit the data. Therefore, skills related to plotting of graphs and finding the best fit can be critical in the success of completing a modelling exercise or task.

Generally, one expects a Level 1 modelling lesson to be trim enough to fit into a one or two-period typical mathematics instructional time.

Level 2

At the next level (Level 2), the focus will be on developing modelling competencies. For the purpose of this discussion, we make a distinction between "competencies" and "skills", the latter having been described above. In contrast, by "competency", we refer to the capacity and capability to apply knowledge specific to mathematical modelling in a modelling problem. Level 2 learning experiences could involve modelling competencies that are inherent in the modelling cycle, such as making assumptions to simplify a problem, identifying factors that influence a dependent variable, or interpreting a mathematical solution in physical terms.

In addition, a Level 2 modelling lesson could be studying and applying known, existing or standard models to a real, physical situation. The related modelling competency here is the ability to recognise the behaviour of a model (or a mathematical function or equation that can potentially be a model) and its use in a real-life problem.

In general, a Level 2 lesson should take up more instructional time than a Level 1 lesson.

Level 3

At the highest level (Level 3), students will be expected to tackle a mathematical modelling task. This should involve substantial complexity and would require students to apply various modelling skills and allow them to further develop their modelling competencies. One may expect students to work in groups, and carry out discussions, develop a model, solve the model and make a presentation.

In other words, Level 3 modelling experiences are what researchers and mathematics educators would have commonly called "modelling tasks". Therefore, one would expect a Level 3 activity to take up quite a fair bit of instructional time, possibly a few hours over a few days.

Having a clear idea of the level to pitch a lesson on modelling is an important step in lesson planning. It provides the teacher with a goal and a structure for the lesson. In addition, it will help the teacher in designing and preparing essential scaffolds for students to successfully complete a task or activity. If students are not ready or sufficiently prepared, the teacher could gradually build their capacity by moving from one level to the next over a period of time.

Mathematical modelling is more than just applying mathematics; it is also about learning mathematics, both concepts and skills, and appreciating what mathematics can do. As pointed out by Warwick (2007), the first step in this learning process is for the student to "become conversant with the tools of the trade" (Warwick, 2007, pp. 34). These include familiarity with mathematical symbols used in the model, algebraic manipulation, computer skills, stages of modelling cycle, and so on. One certainly cannot expect a student with no knowledge of these "tools of the trade" to jump into a modelling task and be successful in carrying out the task.

It is also acknowledged that most modelling tasks are "difficult" for both students and teachers as they are usually cognitively demanding (Blum & Borromeo Ferri, 2009). It is thus unrealistic to expect students (and teachers) to be able to successfully complete a complex modelling task at their first attempt at such tasks. Therefore, it is important that the teacher recognises the need to devise activities to familiarise students with modelling skills, tools and competencies before embarking on more ambitious tasks. For this reason, an auxiliary to the proposed framework is this concept of mapping an idea to the level of modelling experience.

4.4. The Framework

The three levels of learning experiences in mathematical modelling defined in the previous section facilitates a simple framework that teachers can use when planning or designing a lesson on mathematical modelling. The structure of this framework is in the form of a set of five questions listed in Table 4.1.

When planning or designing a modelling lesson, a teacher should first decide *which* level of modelling experience that would be most appropriate

Table 4.1. *Framework for Planning/Designing Mathematical Modelling Learning Experience.*

Framework Component	Explanation
1. WHICH Level of Learning Experience?	Decide which level (Level 1, 2 or 3) of mathematical modelling learning experience that we wish to focus on.
2. WHAT is the Skill/ Competency?	List all the specific skills and competencies (mathematical or modelling) that we target in this learning experience; state the problem to be solved, if applicable.
3. WHERE is the Mathematics?	Write down the mathematical concepts or formulae or equations that will be needed in this learning experience.
4. HOW to Solve the Problem/Model?	Prepare and provide plausible solutions to the problem identified in this learning experience.
5. WHY is this Experience a Success?	List factors or outcomes that can explain why this experience is considered successful and look out for them during the activity.

or suitable that point in time for the students in question. It would not be logical or advisable to plan a, say, Level 3 lesson which requires certain modelling skills or competencies if the target group of students do not have that set of skills or competencies. One could, in fact, "work backwards" and see what is needed in a proposed Level 3 lesson, and plan related Level 2 and Level 1 lessons earlier so that these can serve to support the Level 3 lesson later.

The next question to ask is *what* set of specific skills or competencies that one hopes to develop through this lesson. It will also be good to write down what problem is to be examined or solved in this lesson. It is important to be able to list down these targets as this will eventually help shape the lesson objectives. Doing so will also help the teacher focus the lesson.

The next component in this framework is to specify *where* exactly is mathematics applied, used, taught or practised in this lesson. A mathematical modelling lesson or activity should involve a certain amount and level of mathematics. It would be even better if the teacher, while planning a modelling lesson, is able to link the mathematics found and used in the activity to the syllabus or curriculum, or the teacher's overall scheme of work.

In order to make sure that the modelling activity provides a problem which is actually solvable and manageable by the target group of students, it is crucial that the teacher must have an idea of *how* to approach or even to solve the problem or model. By going through the solution process, the teacher would, in some sense, be experiencing the experience before implementing it. This helps the teacher identify possible blockages and prepare suitable scaffolds.

Finally, the teacher needs to know *why* the lesson planned in this way should ensure success. This last step helps one check and look for success factors, and to have the end in mind. Before implementing the lesson or activity, the expected outcomes are listed. If these outcomes are observed or detected during or after the activity, then one could explain why the lesson is considered a success.

4.5. Applying the Framework

In this section, we present examples to illustrate how the framework proposed in the previous section can be applied. In each example, we begin with an idea, and proceed to use the framework to develop a structure for a modelling lesson or activity or task.

4.5.1. *Example 1: Mountain climbing (Level 1)*

The main idea in this example is to use a data set obtained from the public domain on atmospheric pressure and altitude (see Table 4.2). Such a phenomenon is often experienced when one climbs a mountain and feels the air getting "thinner" at higher altitudes.

The problem could be to find a plausible relationship between the two variables from the data set, and finding an estimate for any parameter that may appear in the model. It turns out that the relationship may be described by an exponential function and students could practise the skill of parameter estimation. Therefore, a Level 1 modelling lesson would be suitable in this case. Using the framework, we could fill up the table as follows.

Framework Component	"Mountain Climbing"
1. WHICH Level of Learning Experience?	Level 1 – about 40 minutes lesson.
2. WHAT is the Skill/Competency?	Knowledge and understanding of the exponential function, and ability to use the Solver Tool in Excel to find parameter value.
3. WHERE is the Mathematics?	Given a set of data, find a function of the form $f(x) = Ae^{-kx}$ that fits the data.
4. HOW to Solve the Problem/Model?	Use of Excel's Solver Tool to find the best value of k that minimises error between data and model.
5. WHY is this Experience a Success?	Students learn a specific skill, and are able to apply it to a problem with real data.

4.5.2. *Example 2: Water warming (Level 2)*

In this example, ice-water is left to warm up and its temperature is recorded at regular intervals. The objective is to study the collected data, make some

Table 4.2. *Atmospheric pressures at different altitudes above sea level.*

Altitude (km)	0	1	2	3	4	5	6	7	8	9	10	11
Pressure (mb)	1013	899	795	701	616	540	472	411	356	307	264	226

sense of it, and then perhaps develop a possible model that would describe how the temperature of the water changes with time. Data may be collected using a temperature probe on a data-logger and recorded via some software (see Figure 4.1).

The problem may be suitably posed as a Level 2 modelling activity in which students are expected to list factors in the problem and make some assumptions about these variables or factors. By inspecting the set of data, one could also propose a possible function that describes how the temperature of the warming water varies with time, and apply a method of estimating any model parameters that may arise from the modelling process.

(a) Temperature probe and a data-logger used to collect data.

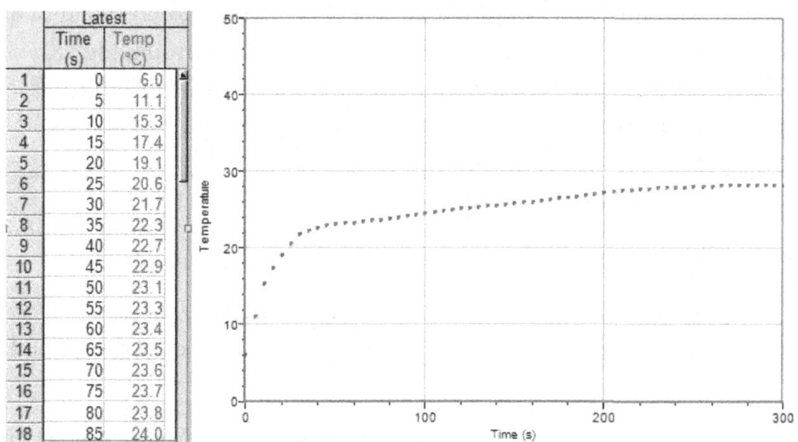

	Latest	
	Time (s)	Temp (°C)
1	0	6.0
2	5	11.1
3	10	15.3
4	15	17.4
5	20	19.1
6	25	20.6
7	30	21.7
8	35	22.3
9	40	22.7
10	45	22.9
11	50	23.1
12	55	23.3
13	60	23.4
14	65	23.5
15	70	23.6
16	75	23.7
17	80	23.8
18	85	24.0

(b) Snapshot of Logger Pro 3 showing variation of temperature of water with time.

Figure 4.1. Recording temperature of ice-water warming up to room temperature.

Alternatively, students may directly apply Newton's law of cooling/warming to explain the observed phenomenon, and complete the solution of the problem by estimating the rate of cooling/warming for this case.

Questions that students may be asked in this activity include:

(1) What are the factors (variables) that can influence or affect the temperature of the water?
(2) What happens near the beginning and end of the experiment?
(3) What assumptions do we need to make about the warming process?
(4) How quickly or slowly does the temperature change at different times?
(5) What can you say about the rate of change of the temperature? Write down a word equation that describes the rate of change.
(6) Write down a differential equation that describes how the temperature changes with time.

Notice that these questions can be handled by students at different cognitive levels. For instance, a pupil in the primary (elementary) school may be able to handle the more basic questions like (1), (2) and (3), whereas a secondary (high school) or junior college (pre university) school student should be able to tackle all the questions, including higher order ones like (4), (5) and (6).

From these questions, hopefully, students can be led to "discover" that the rate of warming (or cooling) of an object is directly proportional to the difference between its temperature (θ) and that of the surrounding (S).

Based on this idea, the basic framework for a Level 2 modelling lesson could be drawn up as follows.

Framework Component	**"Water Warming"**
1. WHICH Level of Learning Experience?	Level 2 — about 1 hour, with homework.
2. WHAT is the Skill/ Competency?	• Listing variables or factors in a model. • Finding suitable equations and estimating model parameters.
3. WHERE is the Mathematics?	• Functions and graphs. • First order differential equations (for students with Calculus background).

(*Continued*)

(Continued)

Framework Component	"Water Warming"
4. HOW to Solve the Problem/Model?	A data-logger is used to collect data. Either: Guess from the behaviour of the points that a possible relationship could be $$\theta = S + (\theta_0 - S)\, e^{kt}$$ where θ_0 is the initial temperature; Or: Apply Newton's Law of Cooling/Warming and the differential equation $$\frac{d\theta}{dt} = k(\theta - S)$$ and use the data-set to estimate the unknown parameter, k.
5. WHY is this Experience a Success?	This lesson helps students to learn to: • identify variables in a real problem, • list assumptions in modelling, • collect and work with data, • make some sense of data, • apply a known model to real data, • apply a method of parameter estimation.

4.5.3. *Example 3: Accident at the MRT station (Level 3)*

A tragic incident happened at one of Singapore's Mass Rapid Transit (MRT) stations in April 2011. As was reported in the local news, a teenage student from Thailand had fallen onto the tracks and was run over by a train approaching the station. She lost both her legs.

In one particular news article, the following was reported (The Straits Times, April 4, 2011).

> *"A 14-year old Thai girl who fell onto an MRT track at the Ang Mo Kio station has lost both her legs. One was severed by a train as it came into the station, and the other was so badly mangled that it had to be amputated. It is not known why Peneakchanasak Nitcharee fell onto the track in the above-ground station at about 11am yesterday, but according to Shin Min Daily News, she had a dizzy spell while waiting on the platform."*

The interesting and perhaps noteworthy point in the report is the claim that the girl could have fallen onto the tracks because "she had a dizzy spell while waiting on the platform". The accident happened at around 11am on a Monday, and one would expect, in a crowded city like Singapore, that there would be some people waiting for the train at that time. Is it possible for a person to be walking randomly (under a "dizzy spell") and falling onto the tracks without being noticed? How many steps would such a random walker take before he/she falls off the platform?

Based on this scenario, a simulation model may be constructed to study and examine the claim made by the girl. One would first need to make some assumptions about the dimensions of the platform, the starting position of the random walk, the rules governing the random walk and so on. The model would then consist of a set of simulation steps, and if possible, these are implemented on the computer. The complexity of the problem makes it suitable as a Level 3 modelling task.

A possible way to handle this problem is to implement a simulation on an Excel spreadsheet. Using integers 1, 2, 3 and 4 to represent four possible directions in which a person can randomly walk, one could construct a simulated random walk and plot the path taken (see Figure 4.2).

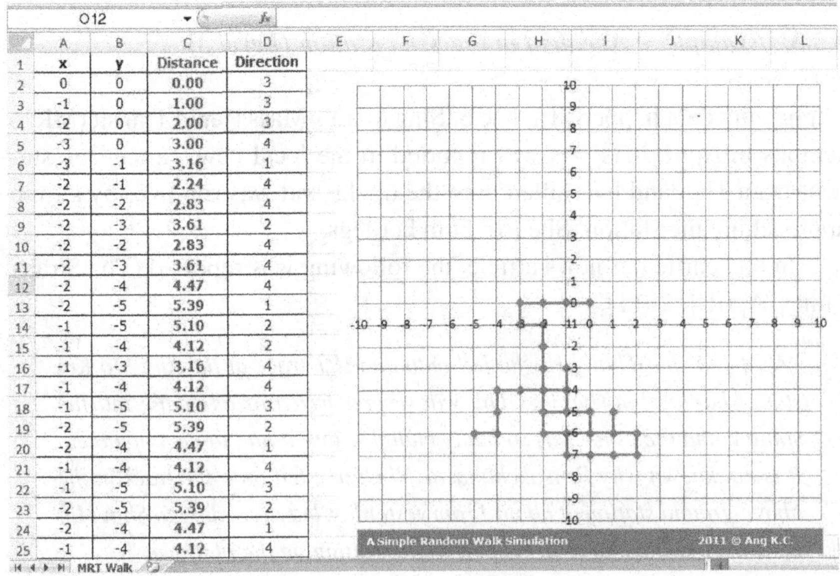

	A	B	C	D
1	x	y	Distance	Direction
2	0	0	0.00	3
3	-1	0	1.00	3
4	-2	0	2.00	3
5	-3	0	3.00	4
6	-3	-1	3.16	1
7	-2	-1	2.24	4
8	-2	-2	2.83	4
9	-2	-3	3.61	2
10	-2	-2	2.83	4
11	-2	-3	3.61	4
12	-2	-4	4.47	4
13	-2	-5	5.39	1
14	-1	-5	5.10	2
15	-1	-4	4.12	2
16	-1	-3	3.16	4
17	-1	-4	4.12	4
18	-1	-5	5.10	3
19	-2	-5	5.39	2
20	-2	-4	4.47	1
21	-1	-4	4.12	4
22	-1	-5	5.10	3
23	-2	-5	5.39	2
24	-2	-4	4.47	1
25	-1	-4	4.12	4

Figure 4.2. A typical run of the simulation on random walk implemented on Microsoft Excel.

This problem can be suitably framed as a Level 3 modelling task for students who have been introduced to probability and the idea of experimenting with chance. Using the framework proposed, we can check the various components as follows.

Framework Component	"MRT Accident"
1. WHICH Level of Learning Experience?	Level 3 — over at least two sessions, possibly with students working in small groups.
2. WHAT is the Skill/Competency?	• Listing variables or factors in a model. • Making assumptions about real physical situation and simplifying problem. • Designing and carrying out a simulation.
3. WHERE is the Mathematics?	• Probability. • Random numbers and the use of a random number generator. • Coordinate geometry.
4. HOW to Solve the Problem/Model?	• Decide on the dimensions of the platform. • Let the walker begin his random walk from the origin. • Generate random numbers 1, 2, 3 and 4 using a computer tool, to represent a step in four different directions. • Compute distance from origin at each step. • Run simulation and see on average how many steps are taken to reach edge of platform.
5. WHY is this Experience a Success?	As a small-group modelling activity, this task allows students to: • work as a group to communicate ideas, • list and agree on suitable and reasonable assumptions, • consolidate conceptual understanding of experimental and theoretical probability (and random numbers), • practise generating random numbers using a computing tool, • design a plausible simulation model.

4.6. Practical Implications

The framework proposed above allows the teacher and curriculum designer to transform an idea or a real situation into a classroom activity on mathematical modelling. Of course, not every idea is suitable, but one could use the framework as a guide to identify the modelling skills or competencies that could be taught, practised or developed.

Another important aspect of the framework is its emphasis on the mathematics needed for the problem solution and the solution process. Apart from getting students to go through the modelling process and cycle, it is also highly desirable for students to learn how to use and apply mathematics to some real-life situations. Using the suggested framework, the teacher is forced to think through the mathematics and the model solution process. One might argue that doing so may result in the teacher focusing on just one solution and stifling students' creativity during implementation. However, as long as the teacher is aware that the mathematics used and solution designed are just one way or one approach to the problem, it should not be a hindrance to the plan. In fact, knowing what mathematics can be applied and how it can be applied would give teachers added confidence when implementing or facilitating the activity.

The hierarchical nature of the framework means that teacher is able to move from one level to the next, depending on the ability and readiness of the students. It also helps the teacher plan a longer term programme or series in mathematical modelling activities, starting with cognitively less demanding lessons at Level 1, and advancing towards more challenging modelling tasks.

It is worth mentioning that when applying the framework, one should keep the modelling experience in focus and not be too enmeshed or carried away in trying to list and unpack the information. The main purpose of the framework is to guide the teacher in developing an idea further and to provide a structure to a mathematical modelling lesson.

4.7. Conclusion

It has been pointed out that while mathematics teachers in Singapore are keen and willing to carry out modelling lessons and activities in their classroom, one of the main challenges facing them is the lack of support and experience in this area (Ang, 2010b). The framework proposed in this discussion provides the teacher with a structured way of selecting an appropriate approach of designing a modelling lesson or activity. This could be an important first step towards developing more sophisticated modelling exercises.

Since the focus of Singapore's mathematics curriculum is on problem solving, some form of modelling may already be taking place in our school classrooms, including in the primary schools. However, these may not be carried out consciously and systematically, as was also observed by Smith (1996) in schools in the United Kingdom. The proposed framework advocates a more systematic and organised way of planning modelling lessons, and provides a practical set of guidelines for the teacher.

It should be noted that the framework does not attempt to include all aspects, elements or components of mathematical modelling in any one particular lesson as this is not the objective. More competencies can be built up over time, and as the teacher becomes more familiar and confident. The framework serves mainly as a way of guiding a novice modeller in moving a possible modelling idea in the real world to a possible modelling lesson or activity in the classroom. In time and with practice, it is hoped that eventually, the teacher will no longer need the framework as thinking through the questions in the framework becomes second nature to the teacher modeller.

References

Ang, K. C. (2001). Teaching mathematical modelling in Singapore schools, *The Mathematics Educator*, 6(1), pp. 63–75.

Ang, K. C. (2004). A simple model for a SARS epidemic, *Teaching Mathematics and Its Applications*, 23(4), pp. 181–188.

Ang, K. C. & Neo, K. S. (2005). A real life application of a simple continuum traffic flow model, *International Journal of Mathematical Education in Science and Technology*, 36(8), pp. 913–956.

Ang, K. C. (2006). Mathematical modelling, Technology and H3 Mathematics, *The Mathematics Educator*, 9(2), pp. 33–47.

Ang, K. C. (2007). Modelling with real data and technology, *Proc. 12th Asian Technology Conference in Mathematics*, pp. 92–102.

Ang, K. C. (2009). *Mathematical Modelling in the Secondary and Junior College Classroom*, (Prentice Hall: Singapore).

Ang, K. C. (2010a). Teaching and learning mathematical modelling with technology, *Proc. 15th Asian Technology Conference in Mathematics*, pp. 19–29.

Ang, K. C. (2010b). Mathematical modelling in the Singapore curriculum: opportunities and challenges, *Proc. Conference on Educational Interfaces between Mathematics and Industry*, pp. 53–61.

Blum, W. & Borromeo Ferri, R. (2009). Mathematical modelling: can it be taught and learnt? *Journal of Mathematical Modelling and Applications*, 1(1), pp. 45–58.

Dindyal, J. (2009). *Applications and Modelling for the Primary Mathematics Classroom*, (Prentice Hall: Singapore).

Galbraith, P. (2006). Real World Problems: Developing Principles of Design, presented at the 29*th Annual Conference of the Mathematics Education Research Group of Australasia*, (RP242006).

Smith, D. N. (1996). Mathematical modelling, *Teaching Mathematics and Its Applications*, 15(1), pp. 37–41.

Stillman, G., Galbraith, P., Brown, J. & Edwards, I. (2007). A Framework for Success in Implementing Mathematical Modelling in the Secondary Classroom, *Proc. 30th Annual Conference of the Mathematics Education Research Group of Australasia*, 2, pp. 688–697.

The Straits Times (2011). "Thai teen loses both legs after being hit by MRT train", news article by Jalelah Abu Baker, http://www.straitstimes.com/Breaking-News/Singapore/Story/STIstory_652713.html (retrieved 10 April 2011).

Warwick, J. (2007). Some reflections on the teaching of mathematical modelling, *The Mathematics Educator*, 17(1), pp. 32– 41.

CHAPTER 5

MATHEMATICAL MODELLING IN AUSTRALIA

Vincent GEIGER

This chapter provides an overview of mathematical modelling in Australia with a particular focus on the state of Queensland. The draft Australian Curriculum: Mathematics is discussed in relation to applications and mathematical modelling. Reflections on what teachers and educators have learned about engaging Australian students in mathematical modelling are presented together with an example of a modelling activity used in a secondary mathematics lesson.

Keywords: Australian Curriculum, Secondary Schools, Mathematical modelling process.

5.1. Introduction

Attention to mathematical modelling within the Australian curriculum can be traced back at least two decades to the seminal Australian document, *A National Statement on Mathematics for Australian Schools* (Australian Education Council, 1990). The purpose of this document was to provide guidelines for all Australian states and territories about what was important for Australian school students to know and be able to do in mathematics. It was intended that the *National Statement* inform state managed curriculum development and influence assessment practices across the nation. The uptake of the advice within the *National Statement* has varied with educational jurisdictions over time to the point where there is now a great deal of variation among states and territories. In this paper an outline is presented of the state of mathematical modelling and applications within Australian schools. In addition, the degree to which it is anticipated that mathematical modelling and applications will be included in the Australian Curriculum: Mathematics will also be discussed. Finally, current practice in mathematical modelling and applications within schools will be illustrated through an example drawn from a secondary school context.

5.2. Mathematical Modelling in the Australian Curriculum

Mathematical modelling is often described as a cyclical process that starts with a problem set in a life-related context which is abstracted into a mathematical representation of the situation and solved through the application of mathematical routines and processes. The solution is then brought into relief against the original problem to consider its fit with the original context. If the fit is not considered sufficient, adjustments are made to the model and the process repeated until a satisfactory fit is achieved. A representation of this process is included in *A National Statement on Mathematics for Australian Schools* (Australian Education Council, 1990), Figure 5.1.

In Australia, the inclusion of mathematical modelling, either as a topic for study or as an assessment practice, is the perrogative of each state or territory based curriculum authority. The uptake by curriculum authorities

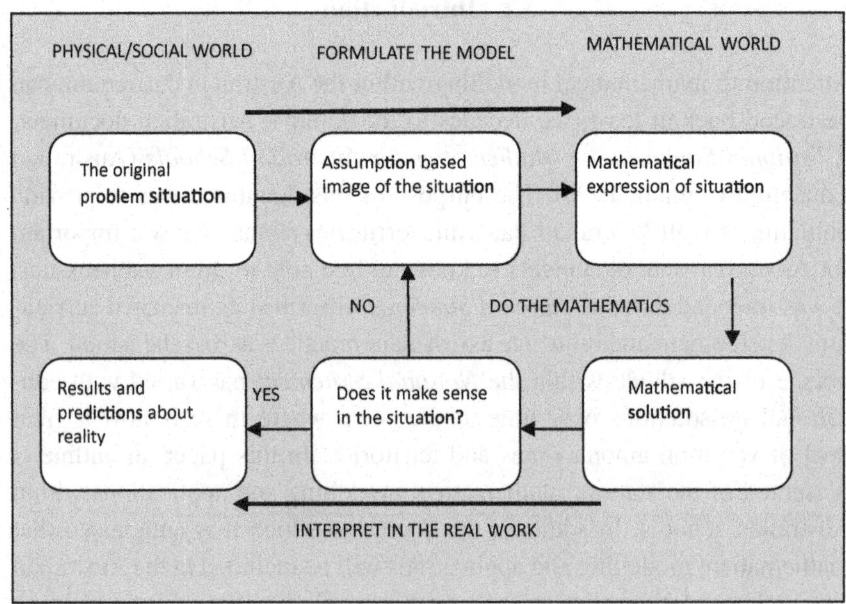

Figure 5.1. Process of mathematical modelling from *A National Statement on Mathematics for Australian Schools* (Australian Education Council, 1990, p. 61).

of mathematical modelling into system wide syllabuses and study designs is a continuum that varies from little or no uptake through to mandatory inclusion in teaching, learning and assessment practices. This continuum will now be illustrated by describing a range of approaches that exist within three different states, New South Wales, Victoria and Queensland. Particular attention will be given to Queensland because of the longevity of mathematical modelling and applications as teaching and learning practice within this state. In order to provide a comparable snapshot of this issue across the nation, a jurisdiction's stance on the importance of mathematical modelling and applications to a student's mathematics education will be inferred from curriculum documents (syllabuses, study designs, etc.) related to what is considered the middle band of mathematics subjects offered to students within their final two years of pre-tertiary study (Years 11 and 12). This selection of documents typically includes studies of function, calculus and, sometimes, introductory statistics.

While New South Wales has a well established reputation for rigour within school mathematics studies, there is little evidence that mathematical modelling receives any serious treatment within mathematics subjects that provide pathways to the study of tertiary courses with a high degree of mathematical demand. A search of the curriculum documents for the Mathematics 2/3 Unit Years 11–12 course (Board of Studies New South Wales, 2008) does not reveal any reference to mathematical modelling.

The Victorian study design for Mathematical Methods (Victorian Curriculum and Assessment Authority, 2010) incorporates the use of Computer Algebra Systems (CAS) into learning and assessment. One of three key outcomes for this course is expressed in the following way.

> *On completion of this unit the student should be able to select and use a computer algebra system and other technology to develop mathematical ideas, produce results and carry out analysis in situations requiring problem-solving, modelling or investigative techniques or approaches.*
>
> (*Victorian Curriculum and Assessment Authority, 2010, p. 30*)

Encouragement to include mathematical modelling and applications is also reflected in advice about the school based component of assessment within Mathematical Methods as schools must draw from the following assessment techniques: assignments; tests; summary or review notes;

projects; short written responses; problem-solving tasks. It should be noted, however, that schools can make a selection from this list and are not required to employ all of the suggested modes of assessment. As a result, while the study and practice of mathematical modelling is encouraged within this Victorian mathematics course it is possible for schools to provide only a light treatment of modelling.

Mathematical modelling has been a mandatory aspect of mathematics in the final years of schooling in all mathematics subjects within Queensland since 1992. While refinements and enhancements have occurred in the interim, mathematical modelling has remained a foundational component of the Queensland Mathematics syllabuses (Stillman and Galbraith, 2009). The current syllabus for Mathematics B (Queensland Studies Authority, 2009), a subject comparable to those from other states discussed above, identifies Modelling and Problem Solving as a General Objective.

> *The objectives of this category involve the uses of mathematics in which the students will model mathematical situations and constructs, solve problems and investigate situations mathematically across the range of subject matter in this syllabus.*
>
> (*Queensland Studies Authority, 2009, p. 4*)

This statement makes it clear that mathematical modelling should be considered as an approach to teaching and learning within a range of content topics. Further, the syllabus states that mathematics needs to be studied and assessed in contexualised situations as well as purely mathematical ones.

> *Students must have the opportunity to recognise the usefulness of mathematics through its application, and the power of mathematics that comes from the capacity to abstract and generalise. Thus students' learning experiences and assessment programmes must include mathematical tasks that demonstrate a balance across the range from life-related to pure abstraction.*
>
> (*Queensland Studies Authority, 2009, p. 6*)

The intent of this syllabus is to instill in students the capacity to solve real-world problems as well as learn the content knowledge and the modes of reasoning associated with mathematics. It is important to note, however, that within Queensland's school-based assessment regime teachers are the generators of mathematical modelling tasks. While this situation could potentially lead to an uneven approach to modelling across the state, panels of experienced teachers monitor standards and provide critical advice to schools at both district and state levels. This process ensures that schools are accountable in relation to the standard of assessment practices associated with the Modelling and Problem Solving assessment criteria.

5.3. The Australian Curriculum: Mathematics

Currently, school curriculums are developed within each Australian state and territory independently. A national curriculum is being introduced in order to foster a greater degree of comparability in student learning opportunities and outcomes across the nation. The initial regime of work being undertaken by the Australian Curriculum and Assessment Reporting Authority (ACARA) has centred on the development of curriculum from pre-school to Year 12 in Mathematics, Science and English. Through 2010 these syllabuses were available in draft form as part of a consultation process, the final curriculum documents are expected to be available in 2011. States and territories will retain responsibility for the implementation of the new curriculum and most seem to have targeted 2011 as a year for teachers to become familiar with the content and requirements of the curriculum with a view to staggered introduction from 2012.

The Pre-school to Year 10 version of the draft National Curriculum provides positive support for the inclusion of applications and mathematical modelling activities in teaching and learning. The rationale for this version of the Australian Curriculum: Mathematics indicates that the use of models is an important approach to the teaching of algebra, for example, through Years 7 to 10.

> *During these years students need to be able to represent numbers in a variety of ways; develop an understanding of the benefits of algebra, through building algebraic models and applications ... explore ways of working with data to allow a variety of representations; and make predictions about events based on their observations.*
>
> (*Australian Curriculum and Assessment Reporting Authority, 2010*)

By contrast, there is less support evident in the first draft of the Years 11 and 12 documents. This curriculum is developed around three subjects. These consist of one subject that focuses on applicable aspects of mathematics that relate to future careers and informed citizenship and two subjects that include studies of function, calculus and statistics that are aimed at students who have aspirations to pursue careers with a high level of mathematical demand. Within the "applicable" mathematics subject, students are expected to engage in investigations which may include mathematical modelling, however, there is currently very little direction within the other two subjects in relation to mathematical modelling and application.

In summary, while there is some encouragement within the P-10 curriculum document of the Australian Curriculum: Mathematics for teachers to include mathematical modelling and applications in their teaching and learning practice, it still remains unclear if this will be articulated into the specifications of the three Years 11 and 12 mathematic subjects.

5.4. A Research Based Example of a Mathematical Modelling Task

There has been significant research activity into the teaching of applications and modelling in Australasia. Studies include investigations into mathematical modelling as curriculum content, strategy selection, task design and teachers' modes of learning about modelling (Stillman, Brown and Galbraith, 2008). Many of these studies have included interesting and engaging examples of the teaching and learning of application and modelling in schools. The episode presented below is chosen as an illustrative example of what modelling and application approaches to teaching and learning mathematics looks like from a secondary mathematics perspective within the Australian context — which in most states incorporates the use of digital tools.

This example is drawn from a study by Geiger, Faragher and Goos (2010) that focused on the role of technology, and especially Computer Algebra Systems, in enhancing the teaching and learning of mathematical modelling in secondary mathematics classrooms. The study explored the collaborative uses of technology by students and investigated the role of technology within each element of the mathematical modelling cycle. The teacher was working with a Year 11 class on a unit about a variety of mathematical functions including linear, quadratic, cubic, exponential and power functions. As part of this unit students were asked to work on the following task.

Table 5.1. *Rate of CO_2 production versus time.*

Time (hrs)	0	1	3	5	7	9	11
Rate of CO_2 (ppm)	0	−0.042	−0.041	−0.038	−0.03	−0.023	−0.008
Time (hrs)	13	15	17	19	21	23	
Rate of CO_2 (ppm)	0.054	0.04	0.03	0.023	0.015	0.005	

Task

The CSIRO has been monitoring the rate at which carbon dioxide is produced in a section of the Darling River. Over a 20-day period they recorded the rate of CO_2 production in the river. The averages of these measurements appear in the table below. The CO_2 concentration [CO_2] of the water is of concern because an excessive difference between the [CO_2] at night and the [CO_2] used during the day through photosynthesis can result in algal blooms which then results in oxygen deprivation and death of the resulting animal population and sunlight deprivation leading to death of the plant life and the subsequent death of that section of the river. From experience it is known that a difference of greater than 5% between the [CO_2] of a water sample at night and the [CO_2] during the day can signal an algal bloom is imminent. Is there cause for concern by the CSIRO researchers? Identify any assumptions and the limitations of your mathematical model.

Students were expected to construct a scatterplot from the provided information and glean the general form of a function that would best fit the data. In turn, this general form needed to be adapted to form a model specific to data in the table. Previously, students were introduced to a technique where ln versus ln plots of data sets were used to determine if a power function was an appropriate basis on which to build a mathematical model. This appears to have influenced the actions of two students as the transcript below indicates.

Researcher: *So do you want to tell me what the plan is?*

Student 1: *The plan is to do the Log/Log plot of both the data to see if they are modelled by a power function. We have previously seen that the . . . .*

Researcher: *So that is something you have learnt to do over time? Whenever you see data look like that, you check if it's a power function by using Log/Log?*

Student 2: *Yes.*

Because students had assumed the data could be modelled by a single power function they encountered a difficulty as the approach meant that students tried to find the natural logarithm of 0.

Student 1: *0.44 zero . . . (entering information into the Nspire device). Don't tell me I have done something wrong. Dammit. Mumbles . . . Start at zero is it possible to do a power aggression* [*sic*]? *I don't think so*!

This comment was in response to the display (Figure 5.2) that resulted when the students attempted to find the natural logarithm of both Time and CO_2 output data using the spreadsheet facility of their handheld device. Students were surprised by the outputs they received for both sets of calculations, that is, the #UNDEF against the 0 entry in the Time column

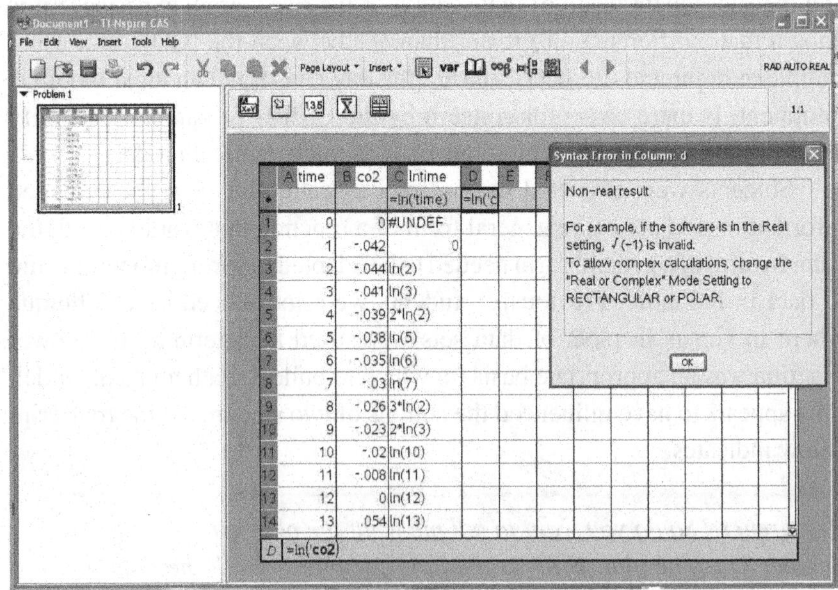

Figure 5.2. Nspire display of spreadsheet for natural logarithm of time and CO_2 output data.

and the lack of any entries in the CO_2 column. After a little more thought students identified the problem with their approach and realised their initial assumption, that is, the best model for the whole data set was a power function, was at fault. This led to an adaptation of their original model in which the data was modelled using two separate functions.

In this excerpt the electronic output forced students to re-evaluate fundamental assumptions they had made in attempting to solve the problem and then to reformulate, solve, interpret and evaluate the problem in the light of an adapted assumption set. Thus, this example indicates that student-student-technology related activity can take place during all phases of the mathematical modelling cycle.

5.5. Conclusion

This chapter has outlined the variation in the implemetation of approaches to mathematical modelling and application in Australian schools. While it is anticipated there will be greater curriculum alignment across states and territories with the advent of the Australian Curriculum: Mathematics, the treatment of modelling and applications within draft National Curriculum documents appears uneven across sectors of schooling and across the proposed senior school subjects.

While it is unclear what focus will be given to applications and modelling in the National Curriculum, there are currently innovative practices already taking place within Australian school classrooms. The example of a modelling task designed for students in the senior phase of schooling demonstrates the way modelling activities can be based around important social or environmental issues. This activity also incorporated the use of digital tools, which allowed for an authentic problem to become accessible to students even though their mathematical knowledge alone may not have allowed them to solve the problem unassisted. This means that a greater range of engaging real-life problems can be presented to students than would be the case if technology was not available. This example demonstrates the opportunities modelling and application approaches provide teachers to enrich and make relevant the teaching of mathematics.

References

Australian Curriculum and Assessment Reporting Authority (2010). *Draft Australian National Curriculum*, 19 March 2010, from http://www.australian-curriculum.edu.au/Learn,

Australian Education Council. (1990). *A National Statement on Mathematics for Australian Schools.* Melbourne: Curriculum Council.

Board of Studies New South Wales (2008). *Mathematics 2/3 Unit Years 11–12.* Sydney: Board of Studies New South Wales.

Geiger, V., Faragher, R. & Goos, M. (2010). CAS-enabled technologies as 'Agents Provocateurs' in Teaching and Learning Mathematical Modelling in Secondary School Classrooms. *Mathematics Education Research Journal*, 22(2), 48–68.

Goos, M. (2007, August). *Developing Numeracy in the Learning Areas (Middle Years).* Keynote address delivered at the South Australian Literacy and Numeracy Expo, Adelaide.

Goos, M., Geiger, V. & Dole, S. (2010). Auditing the Numeracy Demands of the Middle Years Curriculum. In L. Sparrow, B. Kissane & C. Hurst (Eds.), *Shaping the Future of Mathematics Education. Proceedings of the 33rd Annual Conference of the Mathematics Education Research Group of Australasia* (pp. 210–217). Fremantle: MERGA.

Queensland Studies Authority (2009). *Senior Syllabus in Mathematics B.* Brisbane: Queensland Studies Authority.

Stillman, G. & Galbraith, P. (2009). *Softly, Softly: Curriculum Change in Applications and Modelling in Senior Secondary Curriculum in Queensland.* In R. Hunter, B. Bicknell & T. Burgess (Eds.), Crossing divides (Proceedings of the 32nd Annual Conference of the Mathematics Education Research Group of Australasia, pp. 515–522). Palmerston North, NZ: MERGA.

Stillman, G., Brown, J. and Galbraith, P. (2008). Research into the teaching and learning of applications and modelling in Australasia. In H. Forgasz, A. Barkatsas, A. Bishop, B. Clarke, S. Keast, W. Seah & P. Sullivan (Eds.), *Research in Mathematics Education in Australasia, 2004–2007* (pp. 141–164). Rotterdam, The Netherlands: Sense Publications.

Victorian Curriculum and Assessment Authority (2010). *Mathematical Methods.* Melbourne: Victorian Curriculum and Assessment Authority Contents.

CHAPTER 6

MATHEMATICAL MODELLING IN JAPAN

Toshikazu IKEDA

We reviewed the recent trends of teaching modelling in the new Courses of Study and National Achievement Tests and Questionnaires in Japan. Teaching of modelling is one of the central aims in the Courses of Study, but practical aspects of teaching modelling are a concern for many teachers. Second, two typical issues related to the teaching of modelling are discussed: "Why should students solve problems?" and "Can students appreciate A by comparing A with non A?" Third, barriers/obstacles for the inclusion of teaching of modelling in Japan are examined. The fourth issue is how to incorporate open-ended modelling problems into the teaching of mathematical concepts and skills. The roles of open-ended problems are discussed as challenging points from historical perspectives.

Keywords: Japan Curriculum, Obstacles for including Mathematical Modelling in Japanese Schools, Trends of Teaching Modelling in Japan.

6.1. New Courses of Study and National Achieving Tests and Questionnaires

The courses of study are provided as the standard for educational courses in all schools in Japan and the material is revised every 10 years by the Ministry of Education, Culture, Sports, Science and Technology. The current courses of study for compulsory education was published in 2000 and implemented in 2002 under the comprehensive five-day school week system. The current plan was intended to foster "zest for living (IKIRUTIKARA)" and to encourage elements such as robustness, richness of intellect and the ability to independently learn and think.

In mathematics, the number of school hours was reduced by 15% and curriculum contents were reduced by 30%. Further, curriculum content is allotted 80% of total school hours as the minimum for the courses of study and 20% is devoted to both remedial studies and advanced learning.

However after several years, there was a declining trend in the academic abilities of Japanese children, especially in mathematics and science. In the OECD–PISA assessments for mathematical literacy, Japan was in the top position in 2000; however the position dropped to 4th and 6th in 2003 and 2006 respectively. This trend signaled the need for increased emphasis on mathematics and science in the forthcoming revision of the courses of study. New courses of study for the elementary and lower secondary schools were announced in March 2008, and for upper secondary schools the change came in March 2009. The new courses of study will be implemented in 2011 at the elementary school level, in 2012 at the lower secondary school level, and in 2013 in the upper secondary school level. In the new courses of study, time allocation for mathematics is increased compared to the present courses of study as shown in Table 6.1. The number of school hours in a year was increased by 142 for elementary schools and 70 for the lower secondary schools. However, when the number of school hours in future course of study is compared to the previous course of study, the total number is the same. The only difference is the number of schools hours for the 7th grade and 8th grade.

The following four points are emphasised in this new course of study:

(1) More emphasis on Mathematical Activity,
(2) Basic knowledge and skills: Spiral approach,
(3) Mathematical Thinking and Representation,
(4) Meaning and Willingness to Study.

Regarding mathematical activities, applications and modelling is specified as one of the three main activities that should be covered in the classroom. In order to disseminate the spirit of the revised curriculum, national achievement tests and questionnaires were administered to all 6th and 9th grade students and teachers in Japan. Two types of tests were implemented

Table 6.1. *Number of school hours in mathematics.*

Grade	1	2	3	4	5	6	7	8	9
Before	136	175	175	175	175	175	105	140	140
Present	114	155	150	150	150	150	105	105	105
Future	136	175	175	175	175	175	140	105	140

for students, one focused on basic knowledge and skill, and the other targeted applications of mathematics. In the second type of tests, students are presented with problems such as PISA tasks. These tasks test the ability to apply mathematical knowledge and skills in real-life situations and further test the ability to execute, evaluate, and modify a variety of plans to solve a given problem. Disseminating problems, such as PISA tasks for all students at Grades 6 and 9, may be intended to change teachers' belief about the teaching of mathematics. On the questionnaires, elementary and junior high school teachers were asked to respond to the question, "How often do you emphasise the relationship between mathematics and real-world situations?" Possible responses included four alternatives: "often", "sometimes", "infrequently" and "never". For example, the result in Kanagawa which is the second biggest prefecture in Japan was as follows. No significant changes in teachers' treatment of real-world situations in the teaching of mathematics were observed between 2007 and 2008 — for junior high school teachers who responded "often" and "sometimes", 48.5% in 2007 and 49.8% in 2008 were connecting mathematics with the real world. Teaching modelling and connecting mathematics with the real world are not always the same process; therefore we assume that the percentage of teachers who emphasise modelling in the classroom today is even less than previously. Further, we analysed the correlation between the previous questionnaire for teachers and test scores for students. We found that the scores of students whose teachers responded "sometimes" is significantly better than for students whose teachers select "often", "infrequently ", or "never" as shown in Figure 6.1.

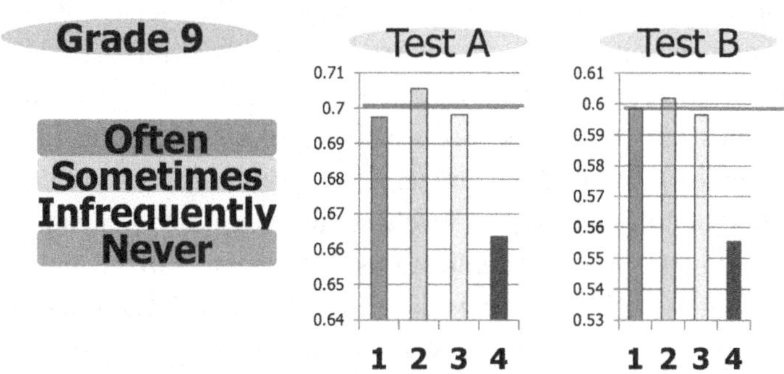

Figure 6.1. Correlation between questionnaires for teachers and test scores for students.

How can these results be interpreted? First, the validity of tests must be examined. Second, the balance between teaching application and modelling and "pure" mathematics is an issue. In junior high school, pure mathematics is important, but balance should be emphasised. Third, the effectiveness of teaching must be examined. Teaching modelling is relatively time consuming. In some cases, teachers might be wasting time by allowing the students to do meaningless activities, so effective teaching is essential.

6.2. Two Typical Issues in Lesson Study Concerning Teaching of Modelling

Lesson study is popular in Japan. Some teachers who emphasise modelling will use lesson study as an effective method to teach mathematics. Two issues regarding modelling are often discussed [Ikeda, 2009].

6.2.1. *Why should students solve problem?*

Students' motivations, which differ according to interests and experiences, should be considered when teaching modelling. Two important points are often discussed. First, what is the reason for solving the problem? Here the background or goals for developing the mathematical model is the issue. It is not always clear to students why they are being asked to solve a particular problem. As a result, it is very difficult to develop a mathematical model that makes sense to every user. Students must understand the reasons before attempting to solve a real-world problem. For example, real-world problems about applying mathematical functions to a given data set are often used in teaching and learning. When the cherry blossoming is forecasted in Japan, this real-world information is important to particular people, because they want to plan a party during the weekend of the bloom. However, the average student is much less interested and therefore less compelled to calculate the cooling rate of coffee. In Japan, teachers are struggling to engage students in appreciating the practical applications that can occur in the real world. In general, a mathematical model is developed to attain a particular purpose, because it is meaningless to apply any function without a purpose. In the case of the cooling rate of coffee, the teacher must frame the appropriate questions to guide the students in their intellectual enquiry, for example,

"How does the temperature decrease as time passes?" and "Is it proportional or not?"

Second, there is always an issue of whether or not students can accept a problem that is posed by someone else, as their own problem. If students can accept the problem as posed, it is an ideal opportunity for the teacher to treat modelling as a classroom activity. One method is to let students select an interesting task from among several modelling tasks presented by the teacher and then the task is completed individually or as a group project. In another method, the teacher proposes a series of observations or actions, and then the students use these to derive similar problems. In this approach, the students must begin by observing or analysing the phenomenon or action. Students are expected to derive key questions which can be solved by using mathematics.

For example, the teacher asks the students, "What shapes of cans are used in a supermarket? Let's examine them this weekend." Through this activity, students discover that most cans are cylindrical though some are in other shapes. Further, considering the relationships between the shape and content of cans, they will find two types. In the first type, the shape is affected by the can's contents, while the second type is not affected by the contents. The students discover that the shapes of cans which are not affected by the contents are generally cylinders. Through these activities, students can gradually formulate a real problem such as, "Why are cans generally shaped like cylinders and not cubes?"

Another example could be how a baton is passed during a relay race [Osawa, 2004]. Students discuss how they can win in a relay race. The students concluded that there are several issues related to winning, such as the order of runners, how to pass the baton, etc. When focusing on the baton pass, the following problem is formulated verbally: "When does the next runner begin to run to get the baton from the previous runner, for the shortest baton pass time? In other words, what is the best distance between the previous runner and the next runner, so that the next runner can get the baton in the shortest possible time?" In this example, the students conceptualise the problem and the necessity for a solution is obvious.

In teaching modelling, it is crucial to propose appropriate observations or actions to stimulate discussion between the teacher and students, so that the students can accept the proposed problem as their own.

6.2.2. *Can students appreciate A by comparing A with non A?*

In teaching modelling, we often discuss the kinds of ideas that are the focus of teaching. After identifying key ideas, we discuss how to implement these ideas in a classroom. Modelling is one of the most effective methods to set up a complex situation that compels students to compare the method of using key ideas versus using an alternative approach that does not use key ideas [Ikeda and Stephens, 2010]. For example, let us consider the situation of "building a mathematical model as simply as possible, and after building the mathematical model gradually modifying it to be more realistic" [Moscardini, 1989]. A question is posed: Is the following sentence true or false? "A mirror, at least half the size of my face, is needed in order to see my whole face [Shimada, 1990; Matsumoto, 2000]". After presenting the problem, the teacher lets the students predict an answer. One student said, "That it might be true because it seems to be half the size of a drawn figure." Another student said, "It might be false because if the mirror is far from my face, it is sufficient to use a small mirror." The teacher directed the students to draw a figure to check their answer. The teacher asked Student A to draw a figure on the blackboard. Figure 6.2. depicts the student's drawing. Making students draw a figure helped them to identify further points that may have been unclear to them.

In fact, the drawing, shown in Figure 6.2, raised several questions from other students, such as

(1) How about the width of the face?
(2) Are three points, namely the eye, the top of the head and the point of chin, on the same plane?

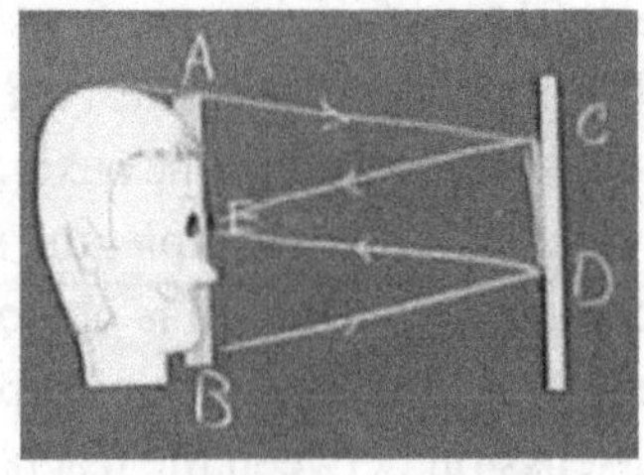

Figure 6.2. Drawing.

(3) Is the relationship of two planes, namely the face and the mirror, parallel or not?
(4) Is the eye located at the midpoint between the top of the head and the point of the chin?

The teacher commended the students' for these comments and questions because they highlight the necessity to clarify vague points in the problem and set up assumptions in order to get a unique solution.

For example, regarding Question 3, two conflicting opinions came from the students. Some students said: "It seems to be easy to solve the problem if the relationship of two planes is parallel." "If the angle is anything except a right angle, it is too difficult to solve the problem." However, another student said: "The relationship of two planes is not always parallel in a real situation." Controversy is expected when the proposed situation is complex. The teacher was able to guide students towards the following idea: "First, let's set up an assumption that the relationship of two planes is parallel. We'll consider the case where the planes are not parallel later on." This teaching scenario demonstrates an instance of "building a mathematical model as simply as possible, and after building the mathematical model gradually modifying it to become more realistic". This example illustrates how students can be taught to compare the method of using key ideas with alternative methods that do not use key ideas. The example also shows the effectiveness of the key ideas of modelling.

6.3. Barriers/Obstacles for the Inclusion of Teaching and Learning Modelling

It is necessary for teachers to consider modelling and mathematical knowledge and techniques when teaching mathematics. When teaching modelling, authentic open-ended problems are required in order to foster modelling competencies, and achieving mathematisation is one of the key stages in teaching modelling. However, in an authentic open-ended problem, there are often too many issues for students to easily discuss and teachers must help different types of students to appropriately organise their thinking. The task of helping students to organise can be difficult and complex, and large amounts of preparation time are required to design lesson plans for modelling. Furthermore, teaching modelling is more time-consuming than teaching mathematics and it is particularly challenging for teachers to ensure that students are mastering specific mathematical concepts and techniques when using authentic open-ended problems.

On the other hand, presenting formulated or mathematical problems makes it relatively easy for teachers to focus on specific issues for discussion and guide students to the intended concepts and techniques. The time required for this type of teaching technique is also shorter than for other methods.

Considering these two approaches, many teachers follow an implicit priority and tend to use formulated or mathematical problems to teach particular mathematical knowledge and techniques; then, if time is available, authentic or open-ended problems are used as a supplement. Current opinion is that students' competencies are more sophisticated thus; it is desirable to change formulaic and closed ways of presenting problems into more open-ended methods so that students can apply their modelling skills.

We should also look at the process of publishing textbooks. In Japan, six publishing companies currently publish elementary and junior high school mathematics textbooks. These companies attempt to include problem situations or teaching processes that match the ideal objectives of modelling in the national curriculum. However, when selecting problem situations or teaching processes, it is also important for publishers to know if teachers are selecting their own textbooks. In order to publish useful material the publishers must ask the teachers for input about what kinds of problem situations or teaching processes are useful in the classroom.

Teachers' responses include the following: modelling is not closely related to entrance examinations; modelling makes teaching more complex and more difficult; and technology that assists with modelling activities is still not readily available. In addition, teachers are apt to mimic teaching methods from their own childhood, so they often select more traditional teaching methods. Thus, authentic or open-ended problems tend not to appear in new textbooks. As a result, intended objectives such as modelling become difficult to implement at a practical teaching level. Niss [2008] noted that the non-negligible "cost" of making applications and modelling a goal in its own right led to a rejection of the goal itself. In this case, the non-negligible "cost" of making a&m a goal in its own right has led to the rejection of authentic open-ended problems in textbooks.

6.4. A Challenging Issue from a Historical Perspective: Roles of Open-Ended Problems

One of the unresolved conflicts in teaching and learning of mathematics is how to strike a balance between modelling and pure mathematics. At the Rome 2008 Centennial of International Commission on Mathematics Instruction, Niss [2008] reiterated the importance of such a balance. Challenging today's accepted opinion that instruction in mathematical ideas and techniques should come *first* and *only then* might it be possible or desirable for students to apply those ideas in modelling activities. As mentioned before, today's student faces an increasingly competitive and sophisticated world and must be capable of addressing open-ended problems and to apply their modelling competencies. We can see a surprising resolution of these conflicts by examining Japanese textbooks [Tyuto Gakko Kyokasyo, 1943] for the junior high school nearly seventy years ago — before and during World War II [Ikeda, 2008; Ikeda, Stephens and Tanaka, 2010].

Historically, Japanese textbook authors tended to use repeated instances of the same concepts through which new phases of mathematical expertise could be developed. This style was identified by Sato [2001] and Tanaka [2008].

By carefully examining the kinds of problems presented in textbooks, an interesting sequence emerges. In the beginning, open-ended problems are posed for students prior to learning any particular mathematical concepts or methods, and then relatively closed problems are posed again using the same context. For example, the open-ended problem in a unit on Measurement for Grade 7 (see Figure 6.3) is posed at first, and two years later, more complex problems are posed in an introduction to the unit "Triangles and Trigonometric function" for Grade 9 as shown in Figure 6.4. After learning trigonometric functions, the same problems are posed again: "*Measure the height of the balloon described in Problem 3 on p. 32 by using trigonometric methods.*"

There is no detailed description of how to treat this open-ended problem (Figure 6.3) in the teacher's guide book; however it is possible to interpret

We can see the rising balloon. Let's devise a method to measure both the height of the balloon and the horizontal distance from here to the point just below the balloon!

Figure 6.3. Unit "Measurement" for Grade 7.

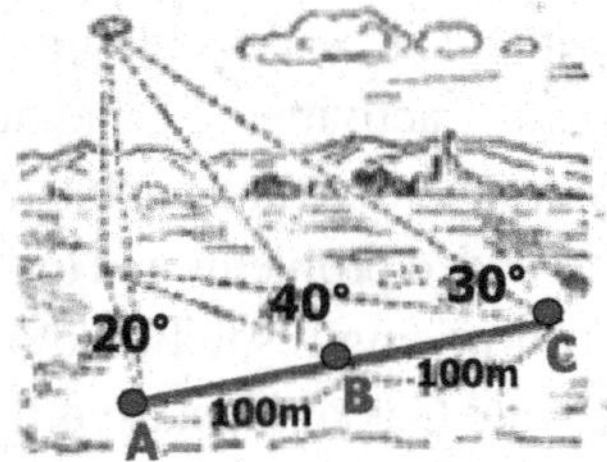

We can see the rising balloon. Select three points, A, B and C, on the same plane with equal distance (100m). The angles between the lines coming from the balloon formed by points A, B and C and the corresponding horizontal lines are 20°, 40° and 30°. Measure the height of balloon by constructing the geometric figure.

Figure 6.4. Introduction of unit "Triangle and Trigonometric Function" for Grade 9.

or reconstruct a teaching sequence which will suggest the role of an open-ended problem before learning a particular mathematical concept. Let's suppose the following three answers (See Figure 6.5) were produced by students, or the teacher elected to present the answers.

These answers are obtained by measuring the distances and angles. Which methods can be used to measure the height of balloon? For ① and ②, it is possible to construct the triangle, and measure/calculate the height of balloon. However, how can we identify the point just below the balloon? Students are expected to recognise that Solution ① is mathematically possible, but finding the point just below the balloon in the real world would be very difficult. There are different issues with ②. The balloon is not static, but will move and many houses or buildings could exist, so it would be difficult to measure the angles from two points to include the point just below the balloon. Next, how about ③?

Students are also expected to find out that the height of balloon is not fixed. Point A cannot be fixed by determining only two angles as seen in

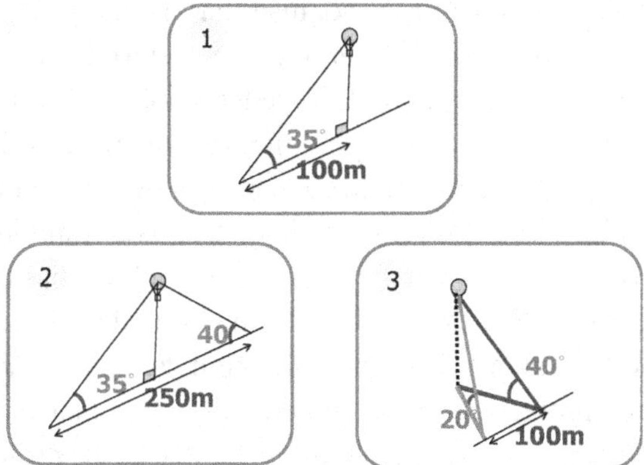

Figure 6.5. Three possible answers for the open-ended problem.

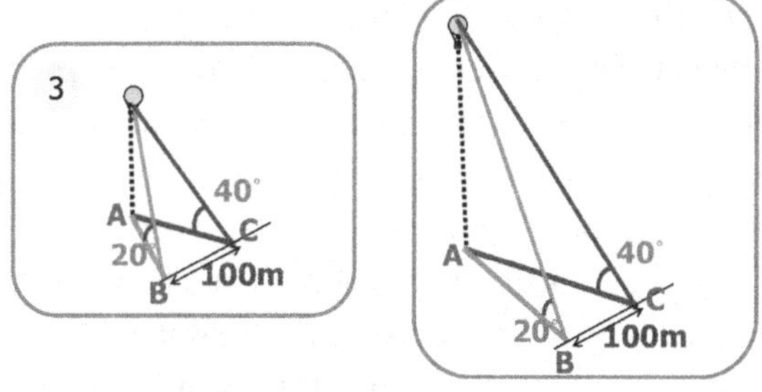

Figure 6.6. Possibility to fixpoint A.

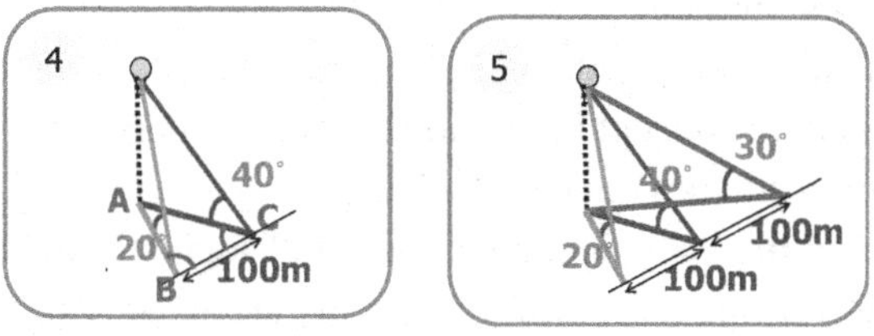

Figure 6.7. Alternative methods to fix the height of balloon.

Figure 6.6. So, the students must make careful modification. Two methods are considered. First, the students could measure an additional two angles as shown in Figure 6.7 (left figure). Second, students can change the method and use three points as shown in Figure 6.7 (right figure).

Students are encouraged to recognise that method ④ is a way to fix point A. In other words, triangulation is used to determine the height of the balloon. How about ⑤? Could Grade 7 students solve this problem? At the Grade 7 knowledge level, it is not possible for students to solve this problem; however Grade 7 students do need to recognise the need for new ideas beyond their current knowledge of mathematics. By examining the possible solution process, we see that the open-ended problem has two important roles for the learners. First, the students come to understand there is more than one method to solve a real-world situation. Second, an important lesson is that some problems can be solved, while others are impossible to solve. Students are also expected to know what makes the problem impossible to solve. They may lack assumptions or the knowledge of conditions in the real-world situation. Verification and modification are necessary as the next step. They may also lack the necessary level of knowledge and skills needed to interpret the problem. Students must eventually realise the necessity to develop new ideas. Furthermore, they have to accept the existence of unsolved problems and learn to temporarily cope until the solution is found.

As the students progress through the different grades levels in school, they will revisit the same formulated problem before and after learning trigonometric geometry. In the next stage, students are expected to learn two additional aspects. First, students come to appreciate that the study of advanced mathematics allows them to solve problems which could not be solved before. Second, they come to understand the characteristics about several sub-systems in mathematics such as direct measurement, elementary geometry and trigonometry.

Posing open-ended problems before the students have studied a particular mathematical concept is a challenge for teachers when we approach modelling in mathematics education. The goal is to help students to realise the importance of mathematical constructs and sophisticated mathematical knowledge and techniques. Additional goals are to provide students with long-term problem solving activities, and the important experience of learning to cope with problems that have yet to be solved.

References

Ikeda, T. (2008). Reaction to M. Niss's plenary talk — Perspectives on the balance between applications and modelling and "pure" mathematics in the teaching and learning of mathematics. In M. Menghini, F. Furinghetti, L. Giacardi, & F. Arzarello (Eds.), *The First Century of the International Commission on Mathematical Instruction (1908–2008) Reflecting and Shaping the World of Mathematics Education* (pp. 85–90). Rome: Encyclopedia Italiana.

Ikeda, T. (2009). Didactical reflections on the teaching of mathematical modelling Suggestions from concepts of "time"and "place", In Morten BlomhØj and Susana Carreira, Roskilde University, Department of Science, Systems and Models, *Mathematical Applications and Modelling in the Teaching and Learning of Mathematics* (pp. 217–228), IMFUFA tekst nr.461.

Ikeda, T. & Stephens, M. (2010). Three teaching principles for fostering students' thinking about modelling: An experimental teaching program for 9th grade students in Japan, *Journal of Mathematical Modelling and Applications*, Vol. 1 No. 2, pp. 49–59.

Ikeda, T., Stephens, M. & Tanaka, Y. (2010). Mathematical modelling giving meaning to mathematics: A case study of Japanese textbooks seventy years ago, *Proceedings of the 5th East Asia Regional Conference on Mathematics Education*, Vol. 2, pp. 591–598.

Matsumoto, S. (2000). Activities that expand the mathematical world through making mathematical models — Considering the length of mirrors that can reflect the whole body, *Journal of Japan Society of Mathematical Education, Mathematical Education*, Vol. 82 No. 1, pp. 10–17 (In Japanese).

Moscardini, A. (1989). The identification and teaching of mathematical modelling skills: In W. Blum, M. Niss and I. Huntley (Eds.), *Modelling, Applications and Applied Problem Solving* (pp. 36–42). Chichester: Ellis Horwood.

Niss, M. (2008). Perspectives on the balance between applications and modelling and 'pure' mathematics in the teaching and learning of mathematics. In M. Menghini, F. Furinghetti, L. Giacardi, & F. Arzarello (Eds.), *The First Century of the International Commission on Mathematical Instruction (1908–2008) Reflecting and Shaping the World of Mathematics Education* (pp. 69–84). Rome: Encyclopedia Italiana.

Osawa, Hironori (2004). Development of applications and modelling by action research in Japanese secondary school. In Hans–Wolfgang Henn and Werner Blum, *ICMI Study 14: Applications and Modelling in Mathematics Education* (pp. 199–204), Study Conference in Dortmund.

Sato, E. (2001). Mathematics Education during World War 2. *The Japanese Journal of Curriculum Studies* Vol. 10, pp. 17–29. (In Japanese).

Shimada, S. (1990). Collected problems for teachers. Tokyo: Kyouritsu Publisher, pp. 41–54 (In Japanese).

Tanaka, Y. (2008). Analysis of teaching materials with common situations in the course, mathematics category, focusing on "mathematization of phenomenon", *Journal of Japan Society of Mathematics Education*, Vol. 90 No. 1, pp. 12–25. (In Japanese).

Tyuto Gakko Kyokasho Kabushiki Kaisya (1943). *Mathematics 1-3 (Textbooks at Grade7-9) Dai Ichirui and Nirui*, Sanseido (In Japanese).

Section 2

Fostering Mathematical Modelling in Schools

CHAPTER 7

LEARNING THROUGH MODELLING IN THE PRIMARY YEARS

Lyn D. ENGLISH

Keywords: Complex Data, Model-Eliciting, Young Children Modelling.

7.1. Introduction

Many nations are highlighting the need for a renaissance in the mathematical sciences as essential to the well-being of all citizens (e.g., Australian Academy of Science, 2006; 2010; The National Academies, 2009). Indeed, the first recommendation of The National Academies' *Rising Above the Storm* (2007) was to vastly improve K–12 science and mathematics education. The subsequent report, *Rising Above the Gathering Storm Two Years Later* (2009), highlighted again the need to target mathematics and science from the earliest years of schooling: "It takes years or decades to build the capability to have a society that depends on science and technology . . . You need to generate the scientists and engineers, starting in elementary and middle school" (p. 9).

Such pleas reflect the rapidly changing nature of problem solving and reasoning needed in today's world, beyond the classroom. As *The National Academies* (2009) reported, "Today the problems are more complex than they were in the 1950s, and more global. They'll require a new educated workforce, one that is more open, collaborative, and cross-disciplinary" (p. 19). The implications for the problem solving experiences we implement in schools are far-reaching. In this chapter, I consider problem solving and modelling in the primary school, beginning with the need to rethink the experiences we provide in the early years. I argue for a greater awareness of the learning potential of young children and the need to provide stimulating learning environments. I then focus on data modelling as a powerful means of advancing children's statistical reasoning abilities, which they increasingly need as they navigate their data-drenched world.

7.2. Rethinking Early Problem Solving

For several decades, mathematics educators have been calling for more relevance and meaning in students' classroom mathematical activities (e.g., Brownell, 1947; Freudenthal, 1991). The need to make mathematics and mathematical problem solving relevant to students' lives has never been greater — nor has been the challenge in doing so. The "real world" of today is becoming rapidly different from that of even five years ago. Advances in technological communications, breakthroughs in medical research, meltdowns in the global economy, and the increasing dominance of complex systems in our lives necessitate a very different approach to "real-world" problem solving in the classroom. In this section, I first consider some issues for children whose problem solving experiences are predominantly one- or two-step word problems. I then argue that a modelling approach can enrich children's problem solving competencies, given that it addresses the mathematical knowledge, processes, representational fluency, and communication skills that they need for the 21st century.

7.2.1. *Word problems*

While acknowledging the importance of "word problems" in developing children's operational competencies, problem solving in the beginning school years needs to extend well beyond a focus on basic operations. In solving these "concept-then-word problems" (Hamilton, 2007, p. 4), children generally engage in a one- or two-step process of mapping problem information onto arithmetic quantities and operations. In most cases, the problem information has already been carefully mathematised for the children (Lesh & Caylor, 2007). Their goal is to unmask the mathematics by mapping the problem information in such a way as to produce an answer using familiar quantities and basic operations. These traditional word problems restrict problem solving contexts to those that often artificially house and highlight the relevant concept (Hamilton, 2007). They thus preclude children from creating their own mathematical constructs.

Students who are fed a diet of stereotyped one- or two-step word problems frequently divorce their real-world knowledge from the solution process, that is, they solve the problems without regard for realistic constraints (e.g., Greer, 1997; Greer, Verschaffel, & Mukhopadhyay, 2007; Selter, 2009; Verschaffel, De Corte & Borghart, 1997). In standard word problems, questions are presented to which the answer is already known by the one asking them (i.e., the teacher). As Verschaffel *et al.* (1997) commented, questions are not given so students can obtain information about an authentic problem situation; rather, the questions are designed to give the teacher information about the students. Furthermore, both the students and the teachers are aware of this state of affairs and act accordingly (cf. Brousseau's, 1997, didactical contract). For example, because students are accustomed to operating on numbers in a word problem to produce an expected numerical answer, they often fail to make sense of the textual representation of the problem and the answer they generate (Selter, 2009).

In contrast, modelling problems shift attention beyond mathematics as computation towards mathematics as conceptualisation, description, and explanation (Lesh, Yoon & Zawojewski, 2007). Such problems require children to make sense of a given situation so that they can mathematise it themselves in ways that are meaningful to them.

7.2.2. *Models and modelling*

The notion of models and modelling has been used variously in the literature, including in reference to solving word problems, conducting mathematical simulations, creating representations of problem situations (including constructing explanations of natural phenomena), and creating internal, psychological representations while solving a particular problem (e.g., English & Halford, 1995; Gravemeijer, 1999; Lesh & Doerr, 2003; Romberg, Carpenter & Kwako, 2005). One perspective on models that I have adopted in my research is that of conceptual systems or tools comprising operations, rules and relationships that can describe, explain, construct, or modify an experience or a complex series of experiences (English, 2011). Modelling involves the crossing of disciplinary boundaries, with an emphasis on the structure of ideas, connected forms of knowledge and the adaptation of complex ideas to new contexts (Hamilton, Lesh, Lester & Brilleslyper, 2008).

Modelling is increasingly recognised as a powerful vehicle for not only promoting students' understanding of a wide range of key mathematical and scientific concepts, but also for helping them appreciate the potential of mathematics as a critical tool for analysing important issues in their lives, communities and society in general (Greer, Verschaffel & Mukhopadhyay, 2007; Romberg *et al.*, 2005). Students' development of powerful models should be regarded as among the most significant goals of mathematics education (Lesh & Sriraman, 2005). Importantly, modelling needs to be integrated within the primary school curriculum and not reserved for the secondary school years and beyond as it has been traditionally. Recent research has shown that primary school children are indeed capable of developing their own models and sense-making systems for dealing with complex problem situations (e.g., English, 2008; English, 2010; English & Watters, 2005).

There is very limited research, however, on developing primary school students' modelling abilities, especially in the early school years. There are a number of possible reasons for this, including a lack of awareness or appreciation of the extent of young children's learning potential and the challenges faced in designing activities in which children generate their own mathematics prior to instruction.

7.3. Young Children's Learning Potential

My research and that of others (e.g., Clarke, Clarke & Cheesman, 2006; English, 2010; English & Watters, 2005; Lesh & English, 2005; Perry & Dockett, 2008) have indicated that young children do possess many conceptual resources which, with appropriately designed and implemented learning experiences, can be bootstrapped toward sophisticated forms of reasoning not typically seen in the early grades. Most research on early mathematics learning has been restricted to an analysis of children's actual developmental level, which has failed to illuminate children's potential for learning under stimulating conditions that challenge their thinking (Ginsburg, Cannon, Eisenband & Pappas, 2006).

More investigations are needed of how young children learn in stimulating environments, in which teachers play a crucial role (Baroody, Lai & Mix, 2006; Ginsburg *et al.*, 2006). "Research on children's current knowledge is not sufficient" (Ginsburg *et al.*, 2006, p. 224). Although useful and important, current research ignores what Vygotsky (1978) referred to as the child's "dynamic mental state" (p. 87) or potential for learning ("zone of proximal development"). As Perry and Dockett (2008) stressed:

> "... young children have access to powerful mathematical ideas and can use these to solve many of the real world and mathematical problems they meet. These children are capable of much more than they are often given credit for by their families and teachers ... The biggest challenge ... is to find ways to utilize the powerful mathematical ideas developed in early childhood as a springboard to even greater mathematical power for these children as they grow older ... (p. 99)."

In heeding Perry and Dockett's advice, consideration needs to be given to the complexity of learning. Winn (2006) warned of the "dangers of simplification", noted that learning is naturally confronted by three forms of complexity — the complexity of the learner, the complexity of the learning material, and the complexity of the learning environment (p. 237). Greater attention is warranted here. In particular, we need to give more recognition to the complex learning that children are capable of — they have greater learning potential than they are often credited by their teachers and families (English, 2010; Lee & Ginsburg, 2007; Perry & Dockett, 2008; *Curious Minds*, 2008). Children have access to a range of powerful ideas and processes and can use these effectively to solve many daily mathematical problems. However, children's mathematical curiosity and talent appear to wane as they progress through school, with current educational practice not cultivating students' capacities (National Research Council, 2005; *Curious Minds*, 2008). It is worth reflecting on the words of Johan van Benthem and Robert Dijkgraaf, the initiators of *Curious Minds* (2008):[1]

> What people say about children is: "They can't do this yet."
> We turn it around and say: "Look, they can already do this."
> And maybe it should be: "They can still do this now."

[1] *Curious Minds* (2008) is a Dutch programme aimed at finding ways to map, preserve, and develop young children's remarkable talents and capabilities regarding the ways they interpret and explain the world around them. The programme addresses concerns that many of these talents seem to disappear on entry to and progress through school. The programme "does not concentrate on static knowledge, but rather on the talents that involve processing such as problem solving, logical reasoning, making an argument, working with multiple representations, estimating chance, classifying, seriating, grasping space and reflecting" processes that "reach far beyond traditional school curricula" (p. 6).

I argue that introducing young children to mathematical modelling where they are challenged to mathematise problematic situations set within appealing, interdisciplinary contexts can pave the way for cultivating their mathematical capacities. Modelling that focuses on reasoning with data, which I term *data modelling* (c.f., Lehrer & Schauble, 2007) provides rich opportunities here.

7.4. Data Modelling in the Early Years

The need to understand and apply statistical reasoning is paramount across all walks of life, evident in the variety of graphs, tables, diagrams and other data representations that need to be interpreted. Primary school children are immersed in our data-driven society, with early access to computer technology and daily exposure to the mass media. With the rate of data proliferation, there have been increased calls for advancing children's statistical reasoning abilities, commencing with the earliest years of schooling (e.g., Franklin & Garfield, 2006; Langrall, Mooney, Nisbet & Jones, 2008; Lehrer & Schauble, 2005; National Council of Teachers of Mathematics [NCTM], 2006; Shaughnessy, 2010). We need to rethink the nature of young children's statistical experiences and consider how we can best develop the important mathematical and scientific ideas and processes that underlie statistical reasoning (Franklin & Garfield, 2006; Langrall *et al.*, 2008; Watson, 2006). One approach in the beginning school years is through data modelling (English, 2010; Lehrer & Schauble, 2007; Lehrer & Schauble, 2000).

Data modelling is a developmental process, beginning with young children's inquiries and investigations of meaningful phenomena, progressing to deciding what is worthy of attention (i.e., identifying attributes of the phenomena), measuring the attributes, and then moving towards organising, structuring, visualising and representing data (Lehrer & Lesh, 2003). Data modelling also involves the fundamental components of beginning inference (Watson, 2006), which include variation and prediction, among others. Figure 7.1 illustrates the cyclic nature of data modelling, where learners would normally cycle iteratively through these components as they undertake an investigation. As one of the major thematic "big ideas" in mathematics and science (Lehrer & Schauble, 2000, 2005), data modelling should

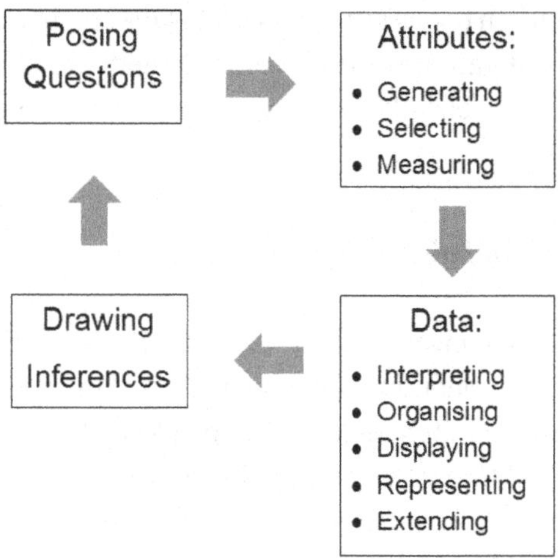

Figure 7.1. Data modelling cycle.

be a fundamental component of early childhood curricula. In the remainder of this section, I address some of these core components of data modelling with young children. I also consider the role of task context.

7.4.1. *Generating and selecting attributes*

Early experiences with data modelling involve selecting attributes and classifying items according to these attributes (Lehrer & Schauble, 2000). As Lehrer and Schauble (2007) noted, it is not a simple matter to identify key attributes for addressing a question of interest — the selection of attributes necessitates "seeing things in a particular way, as a collection of qualities, rather than intact objects" (p. 154). Moreover, children have to decide what is worthy of attention (Hanner, James & Rohlfing, 2002). Some aspects need to be selected and others ignored; the latter of which could be salient perceptually or in some other way. Frequently, however, young children are not given experiences in which they need to consider attributes in this way.

Classification activities presented in the early school years usually involve items with clearly defined and discernable features, such as green square shapes, blue triangular shapes etc. (Hanner *et al.*, 2002). It is thus rather easy for children to classify items of this nature. In contrast, problems involving the consideration of more complex and varied attributes, which could define more than one classification group, present a greater challenge to young children.

7.4.2. *Structuring and displaying data*

Models are typically conveyed as systems of representation, where structuring and displaying data are fundamental — "Structure is constructed, not inherent" (Lehrer & Schauble, 2007, p. 157). However, as Lehrer and Schauble indicated, children frequently have difficulties in imposing structure consistently and often overlook important information that needs to be included in their displays or alternatively, they include redundant information. Providing opportunities for young children to structure and display data in ways they choose, and to critically analyse and assess their representations is important in addressing these early difficulties.

Constructing and displaying their data models involves children in creating their own forms of inscription. By the first grade, children have already developed a wide range of inscriptions, including common drawings, letters, numerical symbols, and other referents. As Lehrer and Schauble (2006) emphasised, developing a repertoire of inscriptions, appreciating their qualities and use, revising and manipulating invented inscriptions and representations, and using these to explain or persuade others, are essential for data modelling. In a similar vein, diSessa (2004) has argued for the development of students' metarepresentational competence, which includes students' abilities to invent or design new representations, explain their creations, and understand the role they play. Yet, students are often taught traditional representational systems as isolated topics at a specified point in the curriculum, without really understanding when and why these systems are used.

7.4.3. *Variation*

Variation lies at the heart of statistical reasoning and is linked to all aspects of statistical investigations (Cobb & Moore, 1997; Garfield & Ben-Zvi, 2007; Watson, 2006). Indeed, as Watson (2006) indicated, the reason data are collected, graphs are created and averages are computed is to "manage variation and draw conclusions in relation to questions based on phenomena that vary" (p. 21).

The importance of variation cannot be underestimated in the development of children's statistical reasoning, beginning with the earliest grade levels (Garfield & Ben-Zvi, 2007). Unfortunately, this is not happening in many classrooms where teachers fail to make specific links to variation whenever they implement activities in data and chance (Watson, 2006). Research on young children's reasoning about variation is limited, although the work of Watson has indicated that young students do have a primitive understanding of variation.

7.4.4. *Role of context*

The nature of task design is a key feature of data modelling activities. Stillman, Brown and Galbraith's (2008) notion of "modelling as vehicle" (p. 143) is applicable here. Such modelling involves choosing contexts in which stimuli for the desired mathematics learning are embedded. Genuine problem situations are used as vehicles for students to construct significant mathematical ideas and processes rather than simply apply previously taught procedures. Furthermore, the mathematics that students engage with in solving such modelling problems usually differs from what they are taught traditionally in the curriculum for their grade level (English, 2003a, 2008; Lesh & Zawojewski, 2007).

When solving data modelling problems, children need to appreciate that data are numbers in context (Langrall, Nisbet, Mooney & Jansem, 2011; Moore, 1990), while at the same time extract the data from the context (Konold & Higgins, 2003). Moore emphasised that a data problem should engage students' knowledge of context so that they can understand and interpret the data rather than just perform arithmetical procedures to solve the problem.

Research has shown that both the data presentation and the context of a task itself have a bearing on the ways students approach problem solution — presentation and context can create both obstacles and supports in developing students' statistical reasoning, emphasising the need to consider carefully task design (e.g., Cooper & Dunne, 2000).

7.5. Study of Data Modelling

In the remainder of this chapter, I report on two activities and selected findings from the second year of a three-year longitudinal study in which we worked with children and their teachers from Grade 1 (2009) through to Grade 3 (2011). The broad aims of the study included tracing and documenting children's development of core statistical ideas and their associated mathematical and scientific learning, and increasing our theoretical and empirical knowledge base by generating paradigms of how new, interdisciplinary data modelling experiences can be implemented within existing curricula to capitalise on young children's potential in handling mathematical and scientific data.

7.5.1. *Participants*

The participants were from an inner-city Australian school. In the first year of the study, three classes of first-grade children (mean age of 6 years 8 months) and their teachers participated. The classes continued into the second year of the study, the focus here (mean age of 7 years 10 months, $n = 68$), and finally, into the third year (mean age of 8 years 8 months). The classroom teachers were vital participants also, as indicated next.

7.5.2. *Design*

A teaching experiment involving multilevel collaboration (English, 2003b; Lesh & Kelly, 2000) was adopted in this study. This approach focuses on the developing knowledge of participants at different levels of learning (student, teacher, researcher) and is concerned with the design and implementation of experiences that maximise learning at each level. The teachers' collaboration in the development of the activities was vital; hence, regular professional development meetings were conducted, as explained below.

The senior research assistant and I met several times with the teachers throughout each year of the study, where we:

(a) reviewed their existing curriculum,
(b) planned an overview of how the new activities would be integrated within their existing curriculum,

(c) collaborated on the development of each of the activities,
(d) discussed children's progress on completion of each activity,
(e) considered refinements for the next activity, and
(f) reviewed the overall achievements at the end of the year, with a focus on both the teachers' and children's developments.

The classroom teachers implemented all of the activities while we observed and collected data, as indicated in Section 7.5.5.

7.5.3. *Interdisciplinary themes and story picture books*

The use of interdisciplinary themes, derived from the teachers' curriculum, is an important feature of the data modelling activities I create with the teachers. In the first year of the study, we chose the theme "looking after our environment". In the second year, the theme was on food selections, with a focus on creating picnics and mini shops.

Another important feature is the use of story picture books, which provide a rich basis for the design and implementation of the activities. Indeed, it is well documented that storytelling provides an effective context for mathematical learning, with children being more motivated to engage in mathematical activities and displaying gains in achievement (e.g., Casey, Kersh & Mercer Young, 2004). In the first year of the study, I created a story picture book about the adventures of Baxter Brown, a "westipoo" — West Highlander X toy poodle. Given that he is one of my dogs and that photos of him appeared in the book, the story had an extra appeal to the children. They asked for more adventures of Baxter Brown in the second year of the study. Hence, the story of *Baxter Brown's Shop* was created, with the accompanying activities, *Baxter Brown's Supermarket Mayhem* and *Baxter Brown's Shop Creation*. The activities were designed within the data modelling framework described in Section 7.4, the core components of which appear in Figure 7.1.

7.5.4. *Activities*

Specifically for the activities addressed here, the aims were to investigate children's "native capacities" (diSessa, 2004, p. 294) to: interpret and

Table 7.1. *Baxter Brown's supermarket mayhem.*

Shopping Trip	Knocked Over	Caught on Shelf	Swished Faces	Covered in Food
Monday	5	3	4	2
Friday	6	2	5	1

analyse tables of data; identify and measure diverse and complex attributes; sort, classify and represent data in a variety of ways; and interpret and analyse diverse representations.

The Baxter Brown's Shop story began with Baxter's mischievous supermarket expeditions undertaken with his owners, Mr and Mrs Brown. The dog generated various forms of mayhem as he raced down the supermarket aisles. Following a whole-class reading of the story, the children were presented the first activity, namely, a simple table of data indicating the different types of mayhem he had created. As a whole class, the children were to determine whether Baxter Brown was becoming more mischievous as his week in the supermarket progressed.

In the second activity, it was explained that Baxter Brown was subsequently banned from the supermarket and thus ended up making his very own shop in his bedroom (the first Baxter Brown story *Baxter Brown's Messy Room* [first year of the study], told of how lucky Baxter Brown was to have his very own bedroom).

For this second activity, the children were given an A3 chart comprising illustrations of 16 supermarket items that displayed diverse attributes including price tags. The items were a carton of milk ($2.47), a frozen pizza ($6.41), apples ($0.90 per kg), coco pops ($7.99), pasta ($2.34), a tin of sliced pineapple ($2.13), fresh carrots ($2.99 per kg), a packet of cheese ($7.86), a packet of bread rolls ($3.00), a packet of biscuits ($4.48), a container of apple juice ($4.09), a carton of eggs ($2.45), a tin of dog food ($1.90), a packet of fish ($4.00), a packaged chicken ($11.96) and a packet of Cheezels ($2.34). Working in small groups, the children responded to a number of written questions that addressed the aims of the activity, including the following:

1. What are some things you notice about the shopping items?
2. What are some ways in which you might sort and classify your data?

After responding to the above two questions, the children were then invited to represent their data, namely, "Now represent your data[2] [in any way you like] on your sheet of paper". On completion, the children were asked, "What are some things that your representation tells about Baxter Brown?"

At the end of the activity, each group presented a report to the class on what they had produced.

7.5.5. *Data collection and analysis*

For the first activity, the three whole-class discussions were videotaped and transcribed. For the second activity, two focus groups in each of the three 2nd grade classrooms were videotaped and audiotaped for in-depth analysis. The range of data collected was analysed using iterative refinement cycles for analysis of children's learning (Lesh & Lehrer, 2000), together with constant comparative strategies (Strauss & Corbin, 1990), where data were coded and examined for patterns and trends

Across the three classes, there were a total of 17 groups of children who completed the second activity, 5 groups in one class and 6 in each of the remaining classes. The analysis of data for Questions 1 and 2 addressed the total number of responses for each question across all groups. For Question 1, there was a total of 41 group responses and for Question 2, there were 24 responses.

7.6. Findings from Activity 1

This activity required children to consider the nature of, and variation in, the given data. The children gave mixed responses to the question, "Do you think Baxter Brown is becoming more mischievous?" There were a few who appeared to weigh the various forms of mayhem and thus responded that Baxter Brown was becoming more mischievous. Comments included: "Yes, because he's knocking stuff off the shelves" and "Yes, because it's just getting worse and worse as it goes down, but there are worse things at

[2]The term, *data*, was familiar to the children from the first year of the study and was also a term included in their curriculum.

the end and better things at the start". In response to the latter comment, one child disagreed stating, "No I think the worse things are at the start and the better things are at the end." In contrast, many children did not consider Baxter Brown was becoming more mischievous as, "The number of accidents is equal." As one child explained, "The numbers on Monday are different to Friday and they're still equal." Other responses included recognition of a pattern in the data, such as "It's in a pattern because first it goes 5 is less than 6, 3 is more than 2, 4 is less than 5, and 2 is more than 1." There was also the use of the story context to justify a decision, such as, Baxter Brown is getting "happier and happier more times, the more times he goes the happier he gets, the more times he wags his tail".

7.7. Selected Findings from Activity 2

7.7.1. *Question 1: Children's identification of attributes*

The children's responses to Question 1 were analysed iteratively with two main categories of responses identified, namely, attributes that were primarily qualitative in nature and those that comprised a quantitative element. Qualitative attributes were favoured over quantitative ones. Of the 41 group responses to this question, 28 were of the former type and 13 of the latter. Examples of the qualitative attributes included "dinner foods, brekky foods, dessert", "foods to cook, foods not to cook", "healthy/non-healthy food" and "cans, bottle, plastic, bags, boxes". Quantitative attributes included, "only one dog food", "all items are under $10 apart from the chicken", "there are only two tins" and "there are more foods than drinks".

7.7.2. *Question 2: Children's identification of ways to sort and classify data*

Half of the 24 group responses to Question 2 referred to complementary categories such as "drinks and food", healthy and not healthy", "cereals and not cereals" and "fruit and non-fruit". 2 groups offered three or more categories, such as "dinner, snack, lunch and breakfast". A further 2 groups suggested putting like items together such as "drinks together, cans together, things that are in boxes together". One other group recommended sorting by cost ("highest price to lowest") and another suggested sorting by shape and size ("big food, small food"; "thickest and thinnest").

7.7.3. *Children's models*

Three main models were evident in the children's recorded representations of their responses to Questions 1 and 2. These were models that comprised (a) lists of items in labelled columns (e.g., Figure 7.2), (b) sets of items enclosed in a curve (e.g., Figure 7.3), and (c) items grouped in two divisions (horizontal or vertical) on the A3 sheet of paper provided (e.g., Figure 7.4).

Rectangular Prism	**Sphere**	**Cone**	**Cylinder**	**Pyramid**
Cheezels (in packet)	Apples	Carrots	Dog food	
Bega (block of cheese)	Chicken	Apple juice	Sliced pineapple	
Coco Pops (in packet)		Fish	Eggs	
Premier (packet of biscuits)				
Pizza (boxed)				
Milk				
Pasta (in packet)				
Buns (in packet)				

Figure 7.2. Example of a model comprising lists of items in labelled columns.

7.7.3.1. Items in columns

5 of the 17 groups developed models that displayed two to five labelled columns, with the names of the items recorded one under the other in the respective columns. One of these models used the shop context to define the categories, namely, "fridge" and "cupboard", while another model listed healthy and unhealthy items with prices attached. The model that listed items in three columns displayed interesting categories, namely, "dry", "wet" and "dry and wet". In their class report, one group member explained, "I said how about things that are dry and things that are wet so we decided to put that down, and then I thought about pasta and I said if you just had pasta by itself it would be dry but if you cooked it, then it would be wet, so then it would be both, so it depends on if you cooked it or if you kept it raw." Another interesting model that comprised five columns displayed items listed under the categories of shapes, namely, "rectangular prism", "sphere", "cone", "cylinder" and "pyramid" (Figure 7.1). The children explained that there were no items that were of a pyramid shape and so that column was left blank.

7.7.3.2. Items enclosed in a curve

Four groups created a model with item names enclosed within a curve. One of these models comprised a large oval, divided into four, with the divisions labelled "healthy", "not healthy", "both" and "dog food" as illustrated in Figure 7.3.

7.7.3.3. Horizontal/vertical division

The most common models created were those that comprised either a vertical or horizontal division of the A3 sheet of paper (either with or without actual dividing lines) and displayed item names or illustrations or both. 9 of the 17 groups developed models of this nature, with one of the groups creating two models, the other being labelled items within a curve. Interesting classifications were evident in these models, such as "combination [of foods]" and "things that taste good by itself". The influence of task context was also visible in these models, such as in the last group who

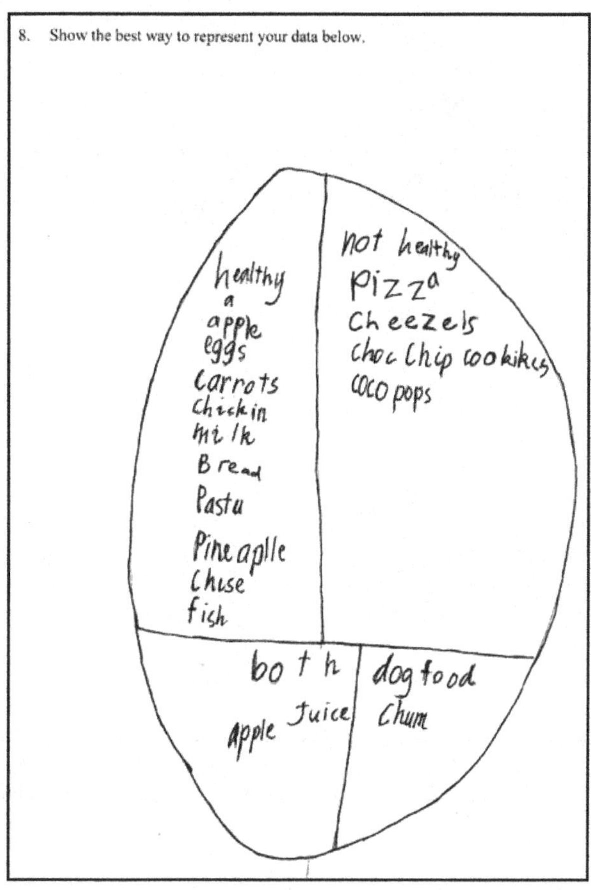

Figure 7.3. Example of a model comprising sets of items enclosed in a curve.

drew two tables (one at the top of the paper and the other at the bottom) with illustrations of items lined up across the tables and prices attached (Figure 7.4).

Another group used a shop context of "fridge" and "cupboard". Their model displayed a drawing of a fridge (labelled "fridge" on the top left-hand corner) with illustrations of cold items stacked on shelves and a cupboard (labelled "cupboard" on the top right-hand corner) with non-cold items illustrated on shelves. Another group incorporated a food pyramid within their model.

Figure 7.4. Example of a model comprising items grouped in two divisions.

7.7.3.4. Children's interpretation of their models

Due mainly to lack of time, not all groups provided responses for the question where they were asked what their representation tells Baxter Brown. Of the 18 group responses, 13 focused primarily on a contextual interpretation (with a common focus on healthy eating) and a literal reading of their model, that is, there was limited interpretation of the data (cf. Curcio, 2010, first level of graph comprehension). Examples of such responses included, "There's a healthy aisle and an unhealthy aisle", "There's dry food and wet food and both" and "It tells him what to put in the fridge and what to put in the cupboard; if he doesn't put it in the right place he might get sick". Only 5 responses referred to any relational observations (cf. Curcio's second level of graph comprehension ["reading between the data"]). Such responses included: "There are more healthy things than unhealthy things", "There is more food than drinks and less drinks than food" and "Our grid tells Baxter Brown what aisles the shape foods are in. It tells him there are no pyramid-shape items like the camera stands tripods. He would have to go to another shop".

7.8. Discussion and Concluding Points

In this chapter, I have argued for the need to reconsider the nature of the problem solving experiences we provide primary school children, especially in the beginning school years where learning potential is high. Data modelling, a powerful approach to advancing children's statistical reasoning abilities, has been the main focus of this chapter.

Data modelling engages children in interpreting and investigating meaningful phenomena involving the identification of diverse and complex attributes; this is in contrast to the standard attributes they usually experience in early mathematics curricula. In building their data models, children organise, structure, visualise and represent their data in ways that they, themselves, choose. Data modelling also involves the fundamental components of beginning inference (Watson, 2006), which include variation and prediction, among others.

Children's responses to Activity 1 indicated that, while they were aware of the variation in the values across Monday and Friday (Table 7.1), the same totals of the two days was the deciding factor for many children. This is an interesting finding, suggesting that these children were not only focusing on the individual values, but were also operating on all the data (that is; totalling the values). Other children took into consideration the nature of the mayhem and appeared to apply a weighting here, such as "knocking stuff off shelves" appeared a worse form of mischief than the others. The story context also played a role in some children's responses, such as Baxter Brown becoming happier as he raced down the shop aisles, with the result that his tail was knocking more items off the shop shelves.

The influence of task context was also evident in children's responses to the second activity, where context appeared to present both support and obstacles in their reasoning. For example, the children's familiarity with the task context appeared to enable them to identify a diverse range of attributes, some quite unexpected, such as a consideration of food combinations, items for different meal times and items that were identified as "dry", "wet" and "dry and wet". On the other hand, the children were very aware of healthy and non-healthy foods (from their health lessons) and this could have overshadowed a possible broader range of attributes being identified. The shop context also influenced some of the groups' model creations where fridges, cupboards, and tables were used to represent data. The impact of task context was further evident in the children's interpretations of their models, where there was a focus on shop aisles and also food storage.

The nature of the task items appeared to have a further impact on the children's identification of attributes and ways to sort and classify the data. Qualitative rather than quantitative features were considered. Nevertheless, the children did identify a wide range of qualitative features despite not making many numerical comparisons. The children's models were not as varied as those created in the first year of the study, (e.g., English, 2010), where the story context focused on Baxter Brown cleaning up his very messy room. Children were given various sets of multiple cut-out items to work with and generated a range of representational models including graphical formats. Perhaps the use of multiple manipulative items might have broadened the children's responses in the second activity. Nevertheless, the children did display an awareness of representational structure in developing their models, with their use of inscriptions enabling clear interpretation and communication of their models. The role of task context in young children's mathematical learning has been highlighted over the years (e.g., Watson, 2006) but this study suggests that further consideration is warranted.

Early years mathematics curricula need to broaden children's problem solving experiences. Data modelling, with its focus on advancing children's statistical reasoning abilities, provides rich opportunities here. However, despite the increased calls for renewed attention to statistical learning in the early school years, research examining young children's statistical developments remains in its infancy.

Acknowledgement

This project reported is supported by a three-year Australian Research Council (ARC) Discovery Grant DP0984178 (2009–2011). Any opinions, findings, and conclusions or recommendations expressed in this paper are those of the author and do not necessarily reflect the views of the ARC. I wish to acknowledge the excellent support provided by my senior research assistant, Jo Macri, and the enthusiastic participation of the children and teachers.

References

Australian Academy of Science (2006). *Critical Skills for Australia's Future.* www.review.ms.unimelb.edu.au (accessed June, 2011).

Australian Academy of Science (2010). Academy 2010 election statement: *Empower Science, Power the Future.* http://www.science.org.au/reports/election-statement.html (accessed August 26, 2010).

Baroody, A. J., Lai, M. & Mix, K. (2006). The development of young children's early number and operation sense and its implications for early childhood education. In B. Spodek & O. Saracho (Eds.), *Handbook of Research on the Education of Young Children* (2nd ed.), Mahwah: Lawrence Erlbaum.

Brousseau, G. (1997). *Theory of Didactical Situations in Mathematics.* Dordrecht: Kluwer.

Brownell, W. A. (1947). The place of meaning in the teaching of arithmetic. *Elementary School Journal*, 47(1), 256–265.

Casey, B. M., Kersh, J. E. & Mercer Young, J. (2004). Storytelling sagas: An effective medium for teaching early childhood mathematics. *Early Childhood Research Quarterly: Special Issue on Mathematics and Science*, 19, 167–172.

Clarke, B., Clarke, D. & Cheesman, J. (2006). The mathematical knowledge and understanding young children bring to school. *Mathematics Education Research Journal*, 18(1), 78–103.

Cobb, G. W. & Moore, D. S. (1997). Mathematics, statistics, and teaching. *American Mathematical Monthly*, 104, 801–823.

Cooper, B. & Dunne, M. (2000). *Assessing Children's Mathematical Knowledge: Social Class, Sex and Problem Solving*. Buckingham, UK: Open University Press.

Curcio, F. (2010). *Developing Data Graph Comprehension* (3rd edn.). Reston, VA: National Council of Teachers of Mathematics.

Curious Minds (2008). The Hague: TalentenKracht.

diSessa, A. A. (2004). Metarepresentation: Native competence and targets for instruction. *Cognition and Instruction*, 22(3), 291–292.

English, L. D. (2003a). Mathematical modelling with young learners. In S. J. Lamon, W. A. Parker & S. K. Houston (Eds.), *Mathematical Modelling: A Way of Life* (pp. 3–18). Chichester, UK: Horwood.

English, L. D. (2003b). Reconciling theory, research, and practice: A models and modelling perspective. *Educational Studies in Mathematics*, 54(2/3), 225–248.

English, L. D. (2008). Introducing complex systems into the mathematics curriculum. *Teaching Children Mathematics*, 15(1), 38–47.

English, L.D. (2010). Young children's early modelling with data. *Mathematics Education Research Journal*, 22(2), 24–47.

English, L. D. (2011, July). Complex modelling in the primary/middle school years. Paper presented at the 15th International Conference on the Teaching of Mathematical Modelling and its Applications.

English, L. D. & Halford, G. S. (1995). *Mathematics Education: Models and Processes*. Mahwah, NJ: Lawrence Erlbaum Associates.

English, L. D. & Watters, J. J. (2005). Mathematical modelling in third-grade classrooms. *Mathematics Education Research Journal*, 16(3), 59–80.

Franklin, C. A. & Garfield, J. (2006). The GAISE project: Developing statistics education guidelines for grades pre-K-12 and college courses. In G. Burrill & P. Elliott (Eds.), *Thinking and Reasoning with Data and Chance* (68th Yearbook, pp. 345–376). Reston, VA: National Council of Teachers of Mathematics.

Freudenthal, H. (1991). *Revisiting Mathematics Education*. China lectures. Dordrecht: Kluwer.

Garfield, J. & Ben-Zvi, D. (2007). How students learn statistics revisited: A current review of research on teaching and learning statistics. *International Statistical Review*, 75(3), 372–396.

Ginsburg, H. P., Cannon, J., Eisenband, J. G. & Pappas, S. (2006). Mathematical thinking and learning. In K. McCartney & D. Phillips (Eds.), *Handbook of Early Child Development* (pp. 208–230). Oxford, England: Blackwell.

Gravemeijer, K. (1999). How emergent models may foster the construction of formal mathematics. *Mathematical Thinking and Learning*, 1, 155–177.

Greer, B. (1997). Modeling reality in mathematics classroom: The case of word problems. *Learning and Instruction*, 7, 293–307.

Greer, B., Verschaffel, L. & Mukhopadhyay, S. (2007). Modelling for life: Mathematics and children's experience. In W. Blum, W. Henn & M. Niss (Eds.), *Applications and Modelling in Mathematics Education: ICMI Study 14* (pp. 89–98). Dordrecht: Kluwer.

Hamilton, E. (2007). What changes are needed in the kind of problem solving situations where mathematical thinking is needed beyond school? In R. A. Lesh, E. Hamilton & J. Kaput (Eds.), *Foundations for the Future in Math Education* (pp. 1–7). Mahwah, NJ: Lawrence Erlbaum.

Hamilton, E., Lesh, R., Lester, F. & Brilleslyper, M. (2008). Model-eliciting activities (MEAs) as a bridge between engineering education research and mathematics education research. *Advances in Engineering Education*, 1(2), 1–25.

Hanner, S., James. E. & Rohlfing, M. (2002). Classification models across grades. In R. Lehrer & L. Schauble (Eds.), *Investigating Real Data in the Classroom* (pp. 99–117). New York: Teachers College Press.

Konold, C. & Higgins, T. L. (2003). Reasoning about data. In Kilpatrick, J., Martin, W. G. & Schifter, D. (Eds.). *A Research Companion to Principles and Standards for School Mathematics*. Reston, VA: National Council of Teachers of Mathematics.

Langrall, C., Mooney, E., Nisbet, S. & Jones, G. (2008). Elementary students' access to powerful mathematical ideas. In L. D. English (Ed.), *Handbook of International Research in Mathematics Education* (2nd ed.) (pp. 109–135). New York: Routledge.

Langrall, C., Nisbet, S., Mooney, E. & Jansem, S. (2011). The role of context expertise when comparing data. *Mathematical Thinking and Learning Journal*, 13(1–2), 47–67.

Lee, J. S. & Ginsburg, H. P. (2007). What is appropriate for mathematics education for four-year-olds? *Journal of Early Childhood Research*, 5(1), 2–31.

Lehrer, R. & Lesh, R. (2003) Mathematical Learning. In W. Reynolds & G. Miller (Eds.) *Comprehensive Handbook of Psychology* (Vol. 7, pp. 357–390). New York: John Wiley.

Lehrer, R. & Schauble, L. (2000). Inventing data structures for representational purposes: Elementary grade students' classification models. *Mathematical Thinking and Learning*, 2(1/2), 51–74.

Lehrer, R. & Schauble, L. (2005). Developing modeling and argument in the elementary grades. In T. Romberg, T. Carpenter & F. Dremock (Eds.), *Understanding Mathematics and Science Matters* (pp. 29–53). Mahwah, NJ: Lawrence Erlbaum Associates.

Lehrer, R. & Schauble, L. (2006). Cultivating model-based reasoning in science education. In R. K. Sawyer (Ed.), *The Cambridge Handbook of the Learning Sciences* (pp. 371–386). NY: Cambridge University Press.

Lehrer, R. & Schauble, L. (2007). Contrasting emerging conceptions of distribution in contexts of error and natural variation. In M. C. Lovett & P. Shah (Eds.), *Thinking with Data* (pp. 149–176). New York: Taylor & Francis.

Lesh, R. & Caylor, B. (2007). Introduction to the special issue: Modeling as application versus modeling as a way to create mathematics. *International Journal of Computers for Mathematical Learning*, 12, 173–194.

Lesh, R. & Doerr, H. (2003). Foundation of a models and modeling perspective on mathematics teaching and learning. In R. A. Lesh & H. Doerr (Eds.), *Beyond Constructivism: A Models and Modeling Perspective on Mathematics Teaching, Learning, and Problem Solving* (pp. 9–34). Mahwah, NJ: Erlbaum.

Lesh, R. & English, L. D. (2005). Trends in the evolution of models & modeling perspectives on mathematical learning and problem solving. *ZDM: The International Journal on Mathematics Education*, 37(6), 487–489.

Lesh, R. A. & Kelly, A. E. (2000). Multi-tiered teaching experiments. In R. A. Lesh & A. Kelly (Eds.), *Handbook of Research Design in Mathematics and Science Education* (pp. 197–230). Hillsdale, NJ: Lawrence Erlbaum.

Lesh, R. & Lehrer, R. (2000). Iterative refinement cycles for videotape analyses of conceptual change. In R. Lesh & A. Kelly (Eds.), *Research Design in Mathematics and Science Education* (pp. 665–708). Hillsdale, NJ: Lawrence Erlbaum.

Lesh, R. & Sriraman, B. (2005). Mathematics education as a design science. *ZDM: The International Journal on Mathematics Education*, 37(6), 490–505.

Lesh, R., Yoon, C. & Zawojewski, J. (2007). John Dewey revisited — Making mathematics practical versus making practice mathematical. In R. A. Lesh, E. Hamilton & J. J. Kaput (Eds), *Foundations for the Future in Mathematics Education* (pp. 315–348). Mahwah, NJ: Lawrence Erlbaum Associates.

Lesh, R. & Zawojewski, J. S. (2007). Problem solving and modeling. In F. Lester (Ed.), *The Second Handbook of Research on Mathematics Teaching and Learning* (pp. 763–804). Charlotte, NC: Information Age Publishing.

Moore, D. S. (1990). Uncertainty. In L. Steen (Ed.). *On the Shoulders of Giants: New Approaches to Numeracy* (pp. 95–137). Washington, DC: National Academy Press.

National Council of Teachers of Mathematics (2006). *Curriculum Focal Points*. Retrieved October 20, 2008, from http://www.nctm.org/standards/content.aspx?id=270.

National Research Council (2005). *Mathematical and Scientific Development in Early Childhood*. Washington, DC: National Academic Press.

Perry, B. & Dockett, S. (2008). Young children's access to powerful mathematical ideas. In L. D. English (Ed.), *Handbook of International Research in Mathematics Education*, 2nd edn. NY: Routledge.

Romberg, T. A., Carpenter, T. P. & Kwako, J. (2005). Standards-based reform and teaching for understanding. In T. A. Romberg, T. P. Carpenter & F. Dremock (Eds.), *Understanding Mathematics and Science Matters*. Mahwah, NJ: Lawrence Erlbaum Associates.

Selter, C. (2009). Stimulating reflection on word problems by means of students' own productions. In L. Verschaffel, B. Greer & S. Mukhopadhyay (Eds.), *Word and Worlds: Modelling Verbal Descriptions of Situations*. Rotterdam: Sense Publishers.

Shaughnessy, M. (2010). *Statistics for All: The Flip Side of Quantitative Reasoning*. Retrieved 14 August, 2010, from http://www.nctm.org/about/content.aspx?id=26327.

Stillman, G., Brown, J. & Galbraith, P. (2008). Research into the teaching and learning of applications and modelling in Australia. In H. Forgasz, A. Barkatsas, A. Bishop, B. Clarke, S. Keast, W. Seah & P. Sullivan (Eds.), *Research in Mathematics Education in Australasia*, 2004–2007 (pp. 141–164). Rotterdam, The Netherlands: Sense Publishers.

Strauss, A. & Corbin, J. (1990). *Basics of Qualitative Research: Ground Theory Procedures and Techniques*. Los Angeles, CA: Sage.

The National Academies (2007). *Rising Above the Storm: Energizing and Employing America for a Brighter Economic Future.* www.national-academies.org (accessed 2/24/2010).

The National Academies (2009). *Rising Above the Gathering Storm: Two Years Later*. http://www.nap.edu/catalog/12537.html (accessed 11/4/2011).

Verschaffel, L., De Corte, E. & Borghart, I. (1997). Pre-service teachers' conceptions and beliefs about the role of real-world knowledge in mathematical modeling of school word problems. *Learning and Instruction*, 7, 339–360.

Vygotsky, L. S. (1978). *Mind in Society*. Cambridge, MA: Harvard University Press.

Watson, J. (2006). *Statistical Literacy at School: Growth and Goals*. Mahwah, NJ: Lawrence Erlbaum.

Winn, W. (2006). System theoretic designs for researching complex events. In J. Elen & R. Clark (Eds.), *Handling Complexity in Learning Environments: Theory and Research* (pp. 238–2540). Elesvier.

CHAPTER 8

FOSTERING MATHEMATICAL MODELLING IN SECONDARY SCHOOLS

Gloria STILLMAN

The change process with any educational innovation does not appear to be something that can be shortened no matter how heightened our awareness of the nature of the change or issues previously met and dealt with in other implementations in other systems according to educational change researchers and theorists. We can, however, benefit from such knowledge of other implementations and the changes teachers underwent by a more in-depth level of implementation in the usual change timeframe. To this end, the change of the Queensland senior secondary mathematics curriculum from a purely mathematical one emphasising content, skills and process to one focusing on applications and mathematical modelling is explored using the voices of teachers reflecting on that change.

Keywords: Queensland Curriculum, Secondary Schools, Teacher Education, Metacognitive Monitoring, Reflecting Learning.

8.1. Introduction

The most extensive and long running implementation of applications and mathematical modelling within a senior secondary curriculum (Years 11 and 12) is in the Australian state of Queensland where both have been a feature of the curriculum for over 20 years (see Stillman & Ng, 2010, for an overview). This implementation is the basis of a longitudinal research investigation, *Curriculum Change in Secondary Mathematics* [CCiSM] (see Stillman & Galbraith, 2009, 2011). Data from interviews conducted for this project will be used to give voice to teachers' views on the changes in practice they have undergone as modelling has gradually taken a place hold within the curriculum.

In an educational system that is entrenched in traditional modes of lesson delivery, it is likely that its school mathematics curriculum would also be traditional. Within such an environment, it would not be easy to make mathematical modelling a part of the mathematics curriculum. (Ang, 2010, p. 53)

Indeed, this was the situation that faced Queensland teachers at a systemic and individual level in the late 1980s and early 1990s as the change from a traditional mathematical curriculum focusing on content, skills and process was first mooted, piloted, trialled and then implemented state-wide.

One might ask from a Singaporean perspective: Why do mathematical modelling in the secondary school mathematics classroom? To begin with, it is in the curriculum as pointed out by Balakrishnan, Yen and Goh (2010):

> Applications and modelling were introduced in the framework of the Singapore mathematics curriculum in 2003 to **highlight their importance in mathematics learning** so as to meet the challenges of the 21st century (Soh, 2005; Ministry of Education, 2009) (p. 247).

However, as one teacher from the long running system-wide implementation of mathematical modelling in the upper secondary school mathematics curriculum in Queensland points out, inclusion in the curriculum is only extrinsic motivation and there are many more reasons mathematics teachers would want to include modelling in schools.

Teacher 9: That is extrinsic motivation but with intrinsic motivation it comes to teachers because if you love your subject you want to make sure that the kids can say, "Hey he taught us some stuff and it was good and it was interesting and I can actually see it in the world around me", so you want to do things like explore the strengths and limitations of a model or make informed decisions.

As Klamkin pointed out many years ago when mathematical modelling began to rise in prominence mainly in Europe, "I have thought for a long time that one of the most important goals of education is to get the students to "think for themselves" (1968, p. 131). Indeed, it was and still is a major goal of education as a second teacher being interviewed with the teacher cited these points out:

Teacher 10: I think we have to teach our kids to think outside the box and that's part of it.

Interviewer: *So you are saying modelling has a spin off into mathematics teaching more generally...*

Teacher 9: *Oh, yes, you know I always loved maths but I don't remember saying to my teachers in the 80s or 70s 'Maths is fun' but now you can't stop saying it because of all these applications. That's what you want to be said in the classroom.*

Teacher 10: *We have got to teach our kids to be researchers and if they don't think, like my son, who is researching, if he hadn't gone outside the box, he wouldn't be able to create. You need that gift to be able to create beyond what's there.*

8.2. The Conundrum for Teacher Educators and Policy Makers

The conundrum for teacher educators and policy makers alike is: "How can teachers be supported to implement mathematical modelling approaches effectively in their classrooms?" Clearly this is not something that can be implemented overnight if long lasting results are to ensue (Stillman, 2007; Stillman & Galbraith, 2009, 2011).

Our tendency as educators or policy writers to describe or name teaching processes in a normative or prescriptive manner is driven by a desire to fully determine the ideal list and combination of teaching techniques which will achieve the goal we desire. Such a view devalues the professional integrity of the teacher from both an institutional and a personal perspective. The complexities of the teaching/learning processes in the mathematics classroom are largely ignored as too is the critical role played by the self efficacy of the teacher as a teacher of mathematics and a teacher of modelling. On the contrary, we have to take all these into account when considering how best to support the teacher as both a teacher of mathematics and a teacher of mathematical modelling.

8.2.1. *Shaping the I teacher*

Inertia or resistance is a natural reaction to new approaches to teaching that require change in practice (Claxton & Carr, 1991; Hall & Hord, 2001). Teachers in the Queensland implementation when it first began felt no different from how teachers in Singapore probably currently feel. Typical responses from interviewees in *CCiSM* to being asked about the existence of any opposition to the original introduction of modelling into the school curriculum were:

Teacher 9: *There was a fair bit of inertia on the part of teachers along the lines: "Where will we get the questions from? How will we do this?"*

Teacher 10: *We did have some in-service training and teachers came out a little bit lost as to what to do. The teachers don't like change usually. . . . Plus also when the people came to us with the package of alternative assessments we couldn't see how they applied to us and how we were going to use them.*

Change in practice requires change in the I as Teacher, that is, "I Teacher" in Figure 8.1. Thus, teachers were expressing personal concerns

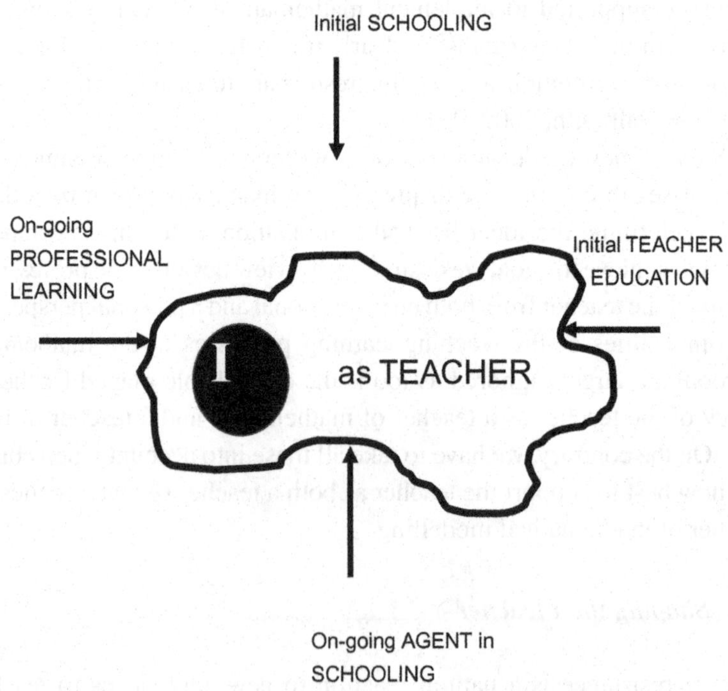

Figure 8.1. Shaping the I teacher.

about how having to incorporate modelling and applications into their teaching would affect them in their classrooms.

To consider some of the issues that arose in the Queensland implementation it could be useful to draw on elements of the Concerns-Based Adoption Model devised by Hall and Hord (2001) for understanding change in educational settings and to facilitate such change effectively. Research literature has confirmed "There is a quasi-developmental path to the concerns as a change process unfolds. However, the flow of concerns is not always guaranteed, nor does it always move in one direction" (Hall & Hord, 2001, p. 63). The expected trajectory according to Hall and Hord, if all supports are in place (i.e., appropriateness of the innovation, principal initiation and careful facilitation), is from personal concerns to task concerns to impact concerns but over a period of 3 to 5 years. The transitioning from personal concerns to task concerns relating to managing the new teaching/learning environment occurs in the first years when the innovation is actually being used. For the case at hand this would be in the first years when modelling would be used in the classroom as opposed to being introduced into curriculum documents. Our ultimate goal in facilitating change is moving teachers to focus on impact concerns where the lens is "on what is happening with students and what the teacher can do to be more effective in improving student outcomes" (Hall & Hord, 2001, p. 59).

Let us begin by looking at the possible reasons for developing personal concerns about incorporating mathematical modelling into secondary teaching. As shown in Figure 8.1, "I Teacher" is somewhat amoebic in outlook with "I" as nucleus and a rather plastic "Teacher" persona. "I Teacher" has been shaped by initial schooling, initial teacher education, the individual teacher's daily carriage of the status of being an educator and an ongoing agent in schooling and any ongoing professional learning undertaken to date. All these influences have moulded "I Teacher" as they are yielded to or pushed aside over time. It is likely that none of these influences on teacher formation included mathematical modelling when change in the curriculum was first mooted. To most teachers in the Queensland implementation when it began, whether they were in schools or teacher education, mathematical modelling and authentic applications were something outside the notion of mathematics taught in the classroom or even personally experienced.

Teacher 3: *When I was teaching . . . supply . . . there was an application of numerical analysis of trying to find the amount of soil in a road excavation and I checked with my husband whether they actually do that and they do. They actually use that sort of method to work out how much soil they have got to move. So I thought, 'That's good. Someone actually uses that. That's not made up by someone.'*

Teacher Educator: *You know how you and I learned matrices at school. I never knew there was a use for them until Maths C came along. This is after doing a university degree in mathematics.*

Personal concerns arise because the challenge is not only to beliefs about the efficacy of modelling in the secondary mathematics classroom but also to beliefs about the nature of mathematics and teachers' efficacy as mathematics teachers. This was evident in the Queensland implementation:

Teacher 9: *The applications are challenging people what the subject is for . . . . We want kids to come out of high school and say, "You can do that with Maths." We want them to say, "Here's something new. You can actually do that with maths . . . "*

The manifestation of personal concerns when an innovation such as modelling is new is thus not a matter for confrontation or dismissal by facilitators of change but a need for empathy, understanding and resolution (Hall & Hord, 2001, p. 77). Clearly, facilitating awareness of the nature of the change to practice, knowledge and beliefs that might be required is essential.

8.2.2. *Assimilating modelling into current practice*

Modelling and problem solving share common features. Problem solving strategies are essential for modelling **but** there is a lot more to the processes of modelling and the meta-knowledge needed to model effectively (cf. amongst others Schukajlow, Leiss, Pekrun, Blum, Müller, & Messner, 2012). Viewing modelling as problem solving or mere application blinds you to the essential features that are distinctive of modelling.

Using mathematical tasks with real world contexts is often claimed to be of interest and enriching for students as their mathematical experiences become connected to the students' world (e.g., Clarke & Roche, 2010). Such sentiments are often expressed by teachers:

Teacher 7: I seem to recall at that time universities were saying there was a decline in interest in mathematics and science, that students were switching to other areas .... And that caused people to say, "Well how can we make mathematics more attractive to the people that are coming through and make it more relevant to what they will do when they go into their tertiary studies?" So I don't think anybody really had to be convinced that we should change our syllabus from pure algebra, pure calculus, pure geometry ... but change the emphasis to applications rather than pure mathematics.

Often such tasks are mathematical applications connecting classroom mathematics to the outside world (Niss, Blum, & Galbraith, 2007, p. 10) rather than the mathematical modelling of real situations outside the classroom. There are important differences between mathematical modelling tasks and mathematical applications.

With a mathematical application the task setter starts with mathematics and reaches out to reality. A teacher designing such a task is effectively asking: "Where can I use this particular piece of mathematical knowledge?" This leads to tasks that illustrate the use of particular mathematics content, a laudable purpose but it is not mathematical modelling. Such tasks are a useful bridge into modelling but are not modelling in themselves as seen in the examples below. In the example shown in Figure 8.2, graphs of functions could be linked to architectural design in the paths of the water spouts from the fountains and the shape of the walls within the water feature at Marina Barrage.

In contrast, with mathematical modelling the task setter begins with reality and looks to mathematics to solve the problem before returning to reality to evaluate the usefulness and suitability of the mathematical model for description or analysis of the real situation of interest. A teacher in designing such a task is thus asking: "Where can I find some mathematics

We are studying graphs of functions.

What in the real world can I use to show the utility of mathematics?

Figure 8.2. An application of functions to the walls at Marina Barrage water feature, Singapore.

to help me with this problem?" In the case of the water feature at Marina Barrage, essentially the task becomes: "I have this problem of how to design a water feature for Marina Barrage so as to fit an available space. How can I use mathematics to model this?"

Being made aware of these distinctions is essential for understanding the fullness of modelling and valuing what makes modelling distinctively different. Without such awareness being explicitly promoted, some teachers see no distinction after many years of incorporating "modelling" into their classroom teaching.

Interviewer: *What about modelling a real life problem as distinct from a mathematical application? Would you see that as a distinction?*

Teacher 5: *Not really, to me modelling a real world problem means you are applying your mathematics so to me the two go together.*

Interviewer: *So you see no difference between a mathematical application and a real world model?*

Teacher 5: *No, no, I don't.*

However, you cannot pick and choose which elements of modelling you will take on board in your teaching as this teacher wants to do, stripping back modelling to mere problem solving.

Teacher 5: *I really would like to see it not so nit picky, not have so many bits and pieces that you have to actually make sure you actually incorporate. I would like to see this trimmed down . . . I am very happy with things like synthesis of procedures and strategies, I think that is very important in the problem solving; selecting appropriate procedures for the particular modelling question involved. It is the exploring the strengths and limitations of models, extending and generalising from solutions, what else? Recognition of the effects of assumptions used, evaluation of the validity of arguments — too much detail there.*

Thus, it is paramount that teachers know what is being advocated as essential to modelling in schooling and what is not.

8.2.2.1. Mathematical modelling elements

The modelling elements being advocated as underpinning the intentions of the Singapore Mathematics Framework (Ministry of Education, 2006) are:

- Mathematisation
- Working with Mathematics
- Interpretation
- Reflection (Balakrishnan, Yen & Goh, 2010, pp. 249–251).

As a first step then these need to be considered as the bottom line for incorporation into teaching practices that effectively incorporate modelling.

8.3. Implementation

Having identified what are essential elements the question then becomes: How can we implement it? Firstly, the educational purposes of incorporating mathematical modelling into the secondary mathematics classroom must be clarified. This will be followed by a brief overview of how modelling activity might be facilitated through metacognitive monitoring.

8.3.1. *Educational purposes of modelling in secondary*

Firstly, modelling needs to be seen as course content in which preserving the integrity of modelling as a mathematical activity is central (Stillman, Brown, & Galbraith, 2010). Secondly, students engage in modelling to develop modelling competencies and meta-knowledge about modelling **and** to obtain a productive outcome for a problem with genuine real world connection (Galbraith, Stillman, Brown, & Edwards, 2007). Thirdly, the solution **must** pay attention to the real world context from which the problem derives (Pollak, 1997). Bearing these three purposes in mind, how teachers foster the metacognitive competencies that are associated with mathematical modelling is crucial to reaching the goal of producing students who can consistently model (Stillman, 2011, p. 169).

8.3.2. *Metacognitive monitoring*

As has been recently confirmed by Magiera and Zawojewski (2011), "social contexts (e.g., small-group modelling sessions) are rich with potential to elicit spontaneous metacognitive activity" (p. 511). During modelling activity the teacher monitors the progress of individuals or groups and intervenes strategically only when necessary. "The teacher can no longer micro-manage the students' thinking or they will not develop the strategic capabilities that are at the centre of modelling" (Antonius, Haines, Jensen, & Niss, 2007, p. 301). To achieve this, careful questioning by the teacher allows the teacher to evaluate the level and quality of metacognitive activities. This could be, for example, the depth with which students are reflecting on the mathematical strategies they have chosen to use or the adequacy of the model(s) they have constructed.

As pointed out in Stillman (2011), teacher reflection on the students' metacognitive activity needs to be both within the specific situation that is being modelled **and** with respect to the role of this metacognitive activity in the modelling process. At the macro level, how a teacher generally undertakes this reflection with respect to student activities and reacts is crucial to whether modelling is nurtured or inhibited in the classroom as a whole despite good intentions with respect to modelling. At the micro level, capacity of students to develop skills associated with modelling, to persist when faced with impasses, and to benefit from both successful and unsuccessful modelling attempts depends on the nature of facilitation in learning and applying the modelling process and metacognitive strategies central to it. Students need to be allowed to think for themselves and modelling is an ideal context in which to do this!

8.3.2.1. Teacher interventions facilitating reflective learning

In order to encourage "reflective learning" (Warwick, 2007, p. 36) where students are continually reflecting on their own learning accommodating new knowledge and new insights, the teacher has to step back from his/her standpoint on the task or deliberately take a neutral position about how the task might be solved (see Doerr & English, 2006, for example). If students have strayed dramatically off task and persisted for some time wallowing in an unproductive state, it is better to try to alter firstly the students' current mental model through their own reflection on this and then the actions of the students rather than the other way around (Warwick, 2007). Sometimes, this is possible through contradiction (see Stillman, 2011, for an extended example). In many instances, however, intervention will be unnecessary when students meet an impasse, as the students themselves are unblocked by peer intervention or in the process of peer discussion as they reflect on their own actions (Stillman, 2011). A perceptive teacher allows time for this to happen. As Doerr points out:

> A modelling approach to teaching mathematics calls for a major reversal in the usual roles of teachers and students. Students need to do more evaluating of their own ideas and teachers need to create opportunities where this evaluation can productively occur" (2007, p. 78).

It is important to realise not all metacognitive acts are productive (see examples in Goos, 1998) so encouraging metacognition for the sake of being metacognitive is not what is being suggested. Some metacognitive acts can result in actions that undermine a perfectly viable solution. Those that are productive involve (a) assessment of personal resources (i.e., knowledge and competence in relation to the task) (b) assessment of viability of alternative strategies and (c) identifying and correcting intermediate errors. Respectively, these lead to recognition that particular strategies are relevant; choosing appropriate strategies for implementation and successful implementation of chosen strategies.

8.4. How Can We Evaluate Modelling?

With any change in curriculum, teachers find there are several concerns that occupy their attention.

Teacher 7: *... modelling and applications were one of the questions that people were asking which was just one of a large sample that accompanies the apprehension that involves any syllabus change. ... see there are a lot of other things claiming people's attention. Assessment was one. How do we assess?*

Modelling processes are mathematical content in their own right and should be assessed. There is no imperative for modelling assessment tasks to incorporate testing of the latest content topic taught. If modelling is seen as an underpinning principle to a curriculum it should not be an optional alternative in assessment. Furthermore, modelling cannot be assessed validly and reliably in timed examination situations (Antonius, 2007; Smith & Thatcher, 1989).

Criteria and standards related to the elements of modelling that are being fostered in the Singapore curriculum (as pointed out in Section 8.2) can be used to assess modelling activity. These can be task specific (see Tanner & Jones, 1993, for examples) or more general assessing how a student has responded across a programme that is underpinned by a modelling approach. If we return to the modelling elements being advocated as underpinning the Singapore Framework, namely Mathematisation (M), Working with Mathematics (WM), Interpretation (I), and Reflection (R), it is seen in Table 8.1 (a set of performance standards developed for the Queensland implementation) that performance at level C or higher is based on the student demonstrating independent modelling. Reflective elements are judged to be distinctive of the highest levels of performance.

Table 8.1. *Standards and performance descriptors for modelling tasks.*

Standard	Description of Performance
A	The overall quality of the students' response generally demonstrates mathematical thinking which includes: • Interpreting, clarifying and analysing, identifying assumptions, parameters and variables in a range of real-world problem situations (M); • Selecting and using effective problem solving and modelling strategies; • Selecting and using mathematical models in a range of real world situations of differing complexity (WM); • Interpreting the results of solving mathematical models in a range of real-world situations of increasing complexity (I); • Articulating the effects of using various assumptions (R); • Articulating the strengths and limitations of models that have been given or developed (R); • Checking the validity of models used (R).
B	The overall quality of the students' response generally demonstrates mathematical thinking which includes: • Interpreting, clarifying and analysing, identifying assumptions, parameters and variables in routine real-world situations and in simple, non-routine real-world situations (M); • Selecting and using effective problem solving and modelling strategies; • Selecting and/or using mathematical models in a range of real-world situations of differing complexity (WM); • Interpreting the results of solving mathematical models in a range of real-world situations of increasing complexity (I);
C	The overall quality of the student's response generally demonstrates mathematical thinking which includes: • Interpreting, clarifying and analysing simple, routine real-world situations (M); • Selecting and/or using mathematical models in simple, routine real-world situations (WM); • Interpreting the results of mathematical activities in simple, routine real-world situations (I).
D	The overall quality of the student's response sometimes demonstrates mathematical thinking which includes: Use of given mathematical models and following of simple problem solving strategies in real world applications
E	The overall quality of the student's response indicates the student implements simple mathematical procedures but is unable to develop basic problem solving strategies.

8.5. Conclusion

In this chapter, how mathematical modelling might be fostered in secondary schools has been explored using the voices of teachers reflecting on the introduction of a now well-established Queensland implementation in a secondary schooling system. Having some knowledge of this implementation is not meant to be a magic formula that will reduce the time for a successful implementation of mathematical modelling in the Singaporean curriculum. On the contrary, as has been revealed in the research and literature on educational change:

> For each new unit (e.g., school, district, or state) that undertakes the change, the process will take three to five years, to a very large extent. For each new adopting unit, the clock begins at the beginning. There are very few shortcuts. (Hall & Hord, 2001, p. 5).

Instead the intention has been to reveal some of the concerns and issues that naturally arise and from an understanding of these help facilitate a higher level of implementation quickly within that time frame. The manifestation of personal concerns during the adoption of an innovation such as modelling can thus be seen as an opportunity to facilitate the change rather than apathy or resistance. By understanding the change process, we have the best chance of facilitating the shaping of "I Teacher" to incorporate mathematical modelling as an on-going part of practice. The author agrees with the following sentiments:

> A mathematics curriculum does not need drastic changes for mathematical modelling to be a part of a learner's learning experience. Indeed, it is more crucial to provide the necessary support that teachers need for them to embrace this mode of teaching mathematics. (Ang, 2010, p. 60).

Thus, how the change process is supported through facilitation will have a crucial impact on its ultimate success and longevity.

References

Ang, K. C. (2010). Mathematical modelling in the Singapore curriculum: Opportunities and challenges. In A. Araújo, A. Fernandes, A. Azevedo, & J. F.

Rodrigues (Eds.), *Conference Proceedings of it EIMI 2010: Educational Interfaces between Mathematics and Industry* (pp. 53–61). Lisbon: Centro Internacional de Matemática & Bedford, MA: COMAP. Available from:http://eimi.glos.org/.

Antonius, S. (2007). Modelling based project examination. In W. Blum, P. L. Galbraith, H-W. Henn, & M. Niss (Eds.), *Modelling and Applications in Mathematics Education* (pp. 409–416). New York: Springer.

Antonius, S., Haines, C., Jensen, T. H., & Niss, M. (with Burkhardt, H.). (2007). Classroom activities and the teacher. In W. Blum, P.L. Galbraith, H-W. Henn, & M. Niss (Eds.), *Modelling and Applications in Mathematics Education* (pp. 295–308). New York: Springer.

Balakrishnan, G., Yen, Y. P., & Goh, L. E. (2010). Mathematical modelling in the Singapore secondary school mathematics curiculum. In B. Kaur & D. Dindyal (Eds.), *Mathematical Applications and Modelling* (pp. 247–257). Singapore: World Scientific.

Clarke, D., & Roche, A. (2010). Teachers' extent of the use of particular task types in mathematics and choices behind that use. In L. Sparrow, B. Kissane, & C. Hirst (Eds.), *Shaping the Future of Mathematics Education* (Vol. 1, pp. 153–160). Adelaide: MERGA.

Claxton, G., & Carr, M. (1991). Understanding change: The personal perspective. In A. Begg *et al.* (Eds.), *SAME papers 1991*. Auckland: Longman Paul.

Doerr, H. (2007). What knowledge do teachers need for teaching mathematics through applications and modelling? In W. Blum, P.L. Galbraith, H-W. Henn, & M. Niss (Eds.), *Modelling and Applications in Mathematics Education* (pp. 69–78). New York: Springer.

Doerr, H. M., & English, L. D. (2006). Middle grade teachers' learning through students' engagement with modelling tasks. *Journal of Mathematics Teacher Education*, 9(1), 5–32.

Goos, M. (1998). 'I don't know if I'm doing it right or I'm doing it wrong!' Unresolved uncertainty in the collaborative learning of mathematics. In C. Kanes, M. Goos, & E. Warren (Eds.), *Teaching Mathematics in New Times* (Vol. 1, pp. 225–232). Gold Coast: MERGA.

Galbraith, Stillman, Brown, & Edwards, (2007). Facilitating middle secondary modelling competencies. In C. Haines, P., Galbraith, W., Blum, & S. Khan (Eds.), *Mathematical Modelling (ICTMA12): Education, Engineering and Economics* (pp. 130–140). Chichester, UK: Horwood.

Hall, G. E., & Hord, S. M. (2001). *Implementing change: Patterns, principles and potholes*. Boston: Allyn & Bacon.

Klamkin, M. S. (1968). On the teaching of mathematics so as to be useful. *Educational Studies in Mathematics*, 1(1), 126–160.

Magiera, M. T., & Zawojewski, J. S. (2011). Characterizations of social-based and self-based contexts associated with students' awareness, evaluation, and regulation of their thinking during small-group mathematical modelling. *Journal for Research in Mathematics Education*, 42(5), 486–520.

Ministry of Education (2009). *The Singapore Model Method for Learning Mathematics*. Singapore: Panpac Education.

Niss, M., Blum, W., & Galbraith, P. (2007). Introduction. In W. Blum, P. L. Galbraith, H-W. Henn, & M. Niss (Eds.), *Modelling and Applications in Mathematics Education* (pp. 3–32). New York: Springer.

Pollak, H.O. (1997). Solving problems in the real world. In L.A. Steen (Ed.), *Why numbers count: Quantitative Literacy for Tomorrow's America* (pp. 91–105). New York: College Entrance Examination Board.

Schukajlow, S., Leiss, D., Pekrun, R., Blum, W., Müller, M., & Messner, R. (2012). Teaching methods for modelling problems and students' task-specific enjoyment, value, interest and self-efficacy expectations. *Educational Studies in Mathematics*, 79(2), 215–237. DOI 10.1007/s10649-011-9341-2.

Smith, R., & Thatcher, D. (1989). An examination for mathematical modelling. *International Journal of Mathematics Education, Science and Technology*, 20(4), 605–613.

Stillman, G. (2007). Implementation case study: Sustaining curriculum change. In W. Blum, P. L. Galbraith, H-W. Henn, & M. Niss (Eds.), *Modelling and Applications in Mathematics Education: The 14th it ICMI study* (pp. 497–502). New York: Springer.

Stillman, G. (2011). Applying metacognitive knowledge and strategies in applications and modelling tasks at secondary school. In G. Kaiser, W. Blum, R. Borromeo Ferri, & G. Stillman (Eds.), *Trends in Teaching and Learning of Mathematical Modelling* (pp. 165–180). New York: Springer.

Stillman, G., Brown, J., & Galbraith, P. (2010). Researching applications and mathematical modelling in mathematics learning and teaching [Editorial]. *Mathematics Education Research Journal*, 22(2) 1–6.

Stillman, G., & Galbraith, P. (2009). Softly, softly: Curriculum change in applications and modelling in the senior secondary curriculum in Queensland.

In R. Hunter, B. Bicknell, & T. Burgess (Eds.), *Crossing Divides* (Vol. 2, pp. 515–522). Adelaide: MERGA.

Stillman, G., & Galbraith, P. (2011). Evolution of applications and modelling in a senior secondary curriculum. In G. Kaiser, W. Blum, R. Borromeo Ferri, & G. Stillman (Eds.), *Trends in Teaching and Learning of Mathematical Modelling* (pp. 689–699). New York: Springer.

Stillman, G., & Ng, D. (2010). The other side of the coin — Attempts to embed authentic real world tasks in the secondary curriculum. In A. Araújo, A. Fernandes, A. Azevedo, & J. F. Rodrigues (Eds.), *Conference Proceedings of it EIMI 2010: Educational Interfaces between Mathematics and Industry* (pp. 491–500). Lisbon: Centro Internacional de Matemática & Bedford, MA: COMAP.

Tanner, H., & Jones, S. (1993). Developing metacognition through peer and self assessment. In T. Breiteig, I. Huntley, & G. Kaiser-Messmer (Eds.), *Teaching and Learning Mathematics in Context* (pp. 228–240). Chichester, UK: Ellis Horwood.

Warwick, J. (2007). Some reflections on the teaching of mathematical modelling. *The Mathematics Teacher*, 17(1), 32–41.

CHAPTER 9

MATHEMATICAL MODELLING — AN EXAMPLE FROM AN INTER-SCHOOL MODELLING CHALLENGE

Raymond BROWN

Trevor REDMOND

Joanne SHEEHY

Dawn LANG

The notion of mathematical modelling is not new; however, with the development and access to cheap, powerful calculator and computer technologies, students in schools are able to engage with sophisticated mathematical contexts that have relevance outside of the classroom, a view of mathematics not commonly held by students. From one of the workshops of modelling challenge, we analyse a conversation between a group of students as a way of providing a meaningful example of the types of conversations students have while they engage in thinking about and doing mathematical modelling.

Keywords: Graphing Calculator, Secondary School, Collective Argumentation, Sociocultural Approach.

9.1. Introduction

A view often held by mathematics teachers is one of students either having or not having the content knowledge of mathematics and then being either able or unable to use it proficiently. Teachers who embrace this view feel that content is fixed and often defined by what is developed in a text book (Smith, p. 390). As a result, the teaching pedagogy adopted by some teachers has been one of 'telling' the students the concepts they should learn. To assist in learning content, the teacher 'tells' the student by showing a worked example, providing little tricks to make attaining the answer easier, and by having the student practise the particular procedure a number of times. If students do not understand the content, "the teacher retells the procedure and the students continue to practise it" (Smith, p. 39).

Occasionally, if time permits; students are exposed to a context which may be construed to be authentic. In these situations, students are given the opportunity to use mathematical concepts and procedures that have been developed to explore that context. In many instances however, the 'authentic contexts' are contrived and the results obtained by students have 'nice' answers. That is, the problem was posed in such a way that there is one right way of arriving at the solution and the solution is unique.

Subsequently, students develop a rigid view of mathematics. That is, they develop a view of Mathematics as having no real use apart from solving problems in a textbook in a classroom setting; that there is a unique way of building a solution to every problem, and that solutions are always 'nice' and found in the back of a textbook (Smith, p. 390). As a result, many students become disengaged with learning mathematics; they do not see mathematics as having any value or use outside the mathematics classroom and they endure their mathematics classes. In turn, this disengagement has an effect on teacher efficacy.

Ashton defines teachers' "sense of efficacy" as a "belief in their ability to have a positive effect on student learning" (Ashton, 1985, p. 142; Smith 1996, p. 388) and that teachers build this sense of efficacy using "perceived past successes"(Smith, p. 389). Given this, the types of experiences the teacher has had in trying different pedagogies will impact not only on their willingness to try something new, but also on their perseverance in the face of adversity. This adversity may come from a number of sources, for example, adversity from students and parents in the way the concepts are being developed, adversity from administration in the form of a lack of provision of resources to adequately ensure the success of the initiative or a lack of professional development or adversity from colleagues that can be manifested in a variety of forms. According to Smith "a sense of efficacy is more appropriately understood as a fluid, dynamic set of beliefs than a fixed personality trait" (p.389). This understanding implies that given the right set of circumstances; appropriate support, adequate professional development and resources, teachers are able to modify and change their sense of efficacy and, in turn, their view of mathematics.

Society values knowledge, and largely sees it as a fluid rather than fixed quantity; however this is not generally attributed to mathematics. Smith (1996) argues that mathematics no longer needs to be seen as a fixed collection of facts and procedures; rather it needs to be seen as a dynamic body of knowledge that is continually enriched through the making of conjectures and through the exploration of analysis and proof. "The fundamental goals of school mathematics are to teach students to understand and reason with mathematical concepts, solve problems arising from new and diverse contexts and develop sense of their mathematical power" (p. 393). Changing the face of mathematics in schools is the 2008 Queensland Mathematics B Syllabus where a number of key competencies are identified and developed on completion of the course. These competencies include "collecting, analysing and organising information, communicating ideas and information, planning and organising activities, working with others and in teams, using mathematical ideas and techniques, solving problems and using technology" (p. 2).

Other sources of expertise such as Walser (2008) identify skill sets necessary for successful citizenship in the 21st Century. These skill sets include abilities related to Critical Thinking, Problem Solving, Collaboration, Written and Oral Communication, Creativity, Self-Direction, Leadership, Adaptability, Responsibility and Global Awareness. The question then becomes, "What may be done to provide teachers with the support necessary to consolidate or refine their sense of efficacy so as to enable them to develop and present courses of study enriched though conjecture and exploration, that will engage students in learning mathematics, and that will promote the development of skill sets that ensure student participation in the 21st Century (Walser, 2008)?" One form of such support may come through engagement in mathematical modelling.

Swertz and Hartzler (1991) suggest that mathematical modelling is "a mathematical process that involves observing a phenomenon, conjecturing relationships, applying mathematical analyses (equations, symbolic structures, etc.), obtaining mathematical results, and reinterpreting the model". For students to be able to build a mathematical model for a particular context, they need to make sense of the context being investigated, identify appropriate assumptions, select the appropriate mathematical concepts and procedures that might be useful in developing a mathematical model and develop a strategy that will allow them to combine those concepts and procedures in meaningful ways to build a model that represents the context. By doing this, students are required to draw on skill sets such as those described by Walser (2008). That is, they are required to be thoughtful as they draw conclusions and develop conjectures by looking for consistencies in the mathematics and consensus between the mathematical model and the context they are attempting to model. If they do not find either the consistency in the mathematics or consensus between the mathematics and the context, then the students are forced to revisit the context, refine or change their assumptions, reconsider the mathematical concepts and strategy they employed and attempt to resolve this conflict.

By engaging in this way, students are given opportunities to realise that the mathematics they develop in the classroom has relevance outside of the classroom. They are also given multiple opportunities to develop and to refine their understanding of the mathematical concepts used to develop the mathematical model. They are also encouraged to question their assumptions. Galbraith and Stillman (2001) have identified that assumptions are developed and refined at different times in the process of building a mathematical model, initially when the student first engages with the context, then during the building phase of the model and finally when the student looks for consensus between the model and the context. Teachers in identifying suitable modelling tasks for students to engage with provide rich social and cultural learning environments for their students centred on teaching and learning.

In sociocultural approachs to teaching and learning, students and expert others work together on tasks to develop solutions to problems that have been posed. It is not the case of the teacher leading the discussion by 'telling' the students the mathematical concepts and then practising those skills many times, nor is it the case of a student blindly blundering on in the hope that at some point in the future they may build a relevant mathematical concept or understanding. Rather, it is a "community of learner's model" (Wilhelm, Baker & Dube, p. 7) where an authentic problem is posed and together, the student and expert other, usually a teacher, but not necessarily so, co-construct a solution for that problem. The teacher intercedes as necessary to ensure the student develops and or adopts the appropriate language to discuss the understandings. This ensures that the necessary mathematical concepts and procedures are being developed, consolidated or refined as necessary so as to allow the student to make progress. The problem needs to be of sufficient complexity to ensure there is challenge in it for the student. However, the complexity should be such that with support the student is able to successfully complete the task (Wilhelm, Baker & Dube, p. 4).

As the teacher works with the student on tasks, the level of support given will vary depending on the ability of the student and the level of interaction provided in the past. As the student and the teacher work together, the student's cognitive level will increase and so the level of support may decrease during successive iterations of the development process (Wilhelm, Baker & Dube, p. 4). One sociocultural approach to teaching and learning that may assist teachers and students to engage with mathematical modelling is Collective Argumentation (Brown & Renshaw, 2000).

9.2. Collective Argumentation

Collective argumentation involves the teacher and students in ways of coming to know, do and value mathematics which reflect the investigative processes and ways of interacting employed by the mathematical community. In simple terms, collective argumentation involves the teacher and students in small group work (2 to 5 students per group) where students are required, initially, to individually "represent" a problem by using pictures, diagrams, drawings, graphs, algorithms, numbers, etc. Students are then required to "compare" their representations with those of other group members. This phase of individual representation and comparison provides the potential for differences in understanding of curriculum content to be exposed and examined. Subsequent talk by the students regarding the appraisal and systematisation of representations is guided by the keywords — "explain", "justify", "agree". Finally, moving from the small group to the classroom collective, the thinking within each group is validated for its consistency and appropriateness as it is presented to the whole class for discussion and validation.

9.2.1. *The teacher's role in collective argumentation*

The teacher has an active role throughout each phase of collective argumentation. The tasks of the teacher include: (a) allocating management of the problem-solving process to the group; (b) facilitating peer co-operation by reminding students of the norms of participation; (c) participating in the development of conjectures and refutations; (d) modelling particular ways of constructing arguments; (e) facilitating class participation in the discussion of the strengths and weaknesses of a group's co-constructed argument, (f) introducing and modelling appropriate mathematical language, and (g) providing strategies for dealing with the interpersonal issues that may arise when working with others.

The remainder of this chapter explores the effects of collective argumentation in making visible students' understandings as they go about knowing and doing mathematics in a novel context — an inter-school mathematics modelling challenge. Specifically, we seek to explore whether a group of students from a collective argumentation classroom see mathematics as providing a forum where personal understanding is privileged, that is, as providing a space where personal understandings can be expressed, re-considered, shared and co-authored.

9.3. The A. B. Paterson College Gold Coast Mathematical Modelling Challenge

The A. B. Paterson College Gold Coast Mathematical Modelling Challenge (The Modelling Challenge) was first conceptualised and implemented in 2004. It is a joint venture between A. B. Paterson College and the Education Faculty of the Gold Coast Campus of Griffith University, supported by numerous academics from various universities, classroom teachers who have a passion for mathematical modelling and sponsored by major corporations.

The Modelling Challenge's initial brief was to cater for students in Years 4 to 11 (that is, 9 to 16 year old students) from schools across South-East Queensland. Invitations were sent to students in these schools with 80 taking up the opportunity to participate in the first year. Over subsequent years, attendance at the Modelling Challenge has continued to grow with 320 students attending in 2009.

In 2009, the Modelling Challenge attracted its first contingent of international students from Singapore. To enhance their experience at the Modelling Challenge, these international students integrated with students from an independent college on the Gold Coast in a Mathematics Modelling Forum which allowed them to explore more of the competencies associated with mathematical modelling.

The drive behind the modelling challenge has always been to provide students with an alternative experience to challenge their thoughts about mathematics instruction and its use beyond the classroom. Many students see mathematics as being only relevant as they work through textbook problems where the answers are known by the teacher and are readily found at the back of the textbook. The modelling challenge aims to provide a significant contrast to this view of mathematics. There are four goals of the modelling challenge.

Goal 1: *To encourage students to operate as mathematicians*: According to the American Mathematical Society's website, "Mathematicians make it possible to send secure emails and buy things online. Mathematicians are essential to analyse data and design accurate models in fields as diverse as biology and finance. Mathematicians enabled researchers to complete The Human Genome Project quickly. And because of the prevalence of the computer at work and at play, mathematicians will continue to touch everyone in modern society." Hence in the modelling challenge, this goal is about encouraging students *to collect data, make propositions, synthesise strategies, and then implement those strategies and assess the validity of the models they produce and the conclusions they derive*. To assist in the realisation of this goal, there is a continuum of task design from the junior year levels through to the senior year levels allowing for different levels of instruction and scaffolding. In the younger years, students are provided with a problem and scaffolding so that they are able to make progress towards a solution. The context is carefully orchestrated so that students have significant opportunity to develop their own conjectures, test those conjectures and then to refine them. In Years 10 and 11, students identify a context and pose a problem of interest to them based on that context, collect data, and develop mathematical models and use those models to make predictions, all the while, identifying the assumptions necessary for their models to be valid and determining the limitations associated with their model and the process. Across all the year levels, students learn about the modelling cycle while they engage with the context and are provided feedback by a mentor. The students are encouraged to engage in multiple iterations of the modelling cycle.

Goal 2: *To provide opportunities for students to engage in tasks that are novel and interesting, where the answer is not readily accessible by either the teacher or the student and where the responses are dependent on the assumptions the students make.* In the junior classes, students are generally involved in building something concrete from which they are able to collect data. In past modelling challenges, students have considered problem contexts such as 'What shaped rotor gives the greatest energy output for a wind generator?'. Students were given the generator, and materials to make a rotor for a wind generator. They then needed to determine the area the rotor and measure the energy generated by that particular configuration using a voltage probe. Students subsequently reduced the area of the rotor and considered the effect this had on the amount of energy produced. When the rotor had less area, it could spin faster thus generating more electricity, however, if the area became too small, it would not receive sufficient thrust from the wind and so electricity generation would reduce. Students were required to determine (for their rotor design) which configuration had the optimum electricity generation.

Goal 3: *To work with students from different schools.* When mathematicians work together to solve a problem they bring their individual strengths and experiences to the table to illuminate the problem and postulate different strategies. In many instances, they may only work together on this project and then disperse; we want students to experience this. To facilitate this goal, students are placed into groups of four within the year levels: Years 4 and 5, Years 6 and 7, Years 8 and 9, and Years 10 and 11. Ideally, the groups are constructed so students from different schools are spread across the groups with care taken to ensure there is a balance across the year levels and the gender of the students. Groups in the upper age levels tend to be more fluid and while they are initially placed into groups, group structure tends to be more fluid. Hence in the modelling challenge it is the intention of this goal to encourage students to work collaboratively to model a real, life situation.

Goal 4: *To provide students with sufficient time to be able to try different ideas and ascertain "Which solution is better?"* In many classroom situations, students get the idea that there is only one single answer and one single procedure used to get that answer and that the teacher has all the answers. They may also ascribe to the belief that to be good at mathematics, they must use a specific procedure to arrive at a specific answer. The modelling challenge attempts to facilitate in student's thinking, a different view of mathematics; one where time is needed, where they are able to engage with the task, to identify the mathematics they know that they feel will be useful, to create ideas, and to reflect on the conclusions they have made. Hence the modelling challenge is conducted over two days. The first day allows students to engage with the task and to try different strategies. During the second day the students, then armed with the results of the previous day, are encouraged to reflect upon and to refine their strategies and the resulting predictions. Each group prepares a poster of their findings and makes an oral presentation of their work to their peers and the class mentor. During this presentation, students are able to question and to challenge the conclusion provided by the presenting group. This provides more feedback to the students in terms of their conclusions and can challenge them to think more deeply about their conclusions and the mathematics used to develop those conclusions.

9.4. A Case Study

9.4.1. *Participants*

This case study will follow the conversations of a group of students as they engage in a particular modelling task. This group of students participated in the Year 6/7 section of the inter-school mathematics modelling challenge. Three of these students (two girls and one boy) — Helen, Nicole, and Neil — form the focus of this study.

9.4.2. *The task*

Each group was required to design a mini-golf hole — complete with blockers, tunnels and other obstacles — and create a theoretical hole-in-one path of the ball such that each angle of incidence equalled the corresponding angle of reflection. Over the two days of the modelling challenge, the students were engaged with the task of designing, building to scale, and modelling a mini-golf hole mathematically. Initially each group was required to represent their mini-golf hole design on graph paper, provide a spreadsheet showing the segment angles and linear equations, and provide a short journal entry of their experience with the challenge. Each group received a poster board, a piece of green felt, wooden blocks, cardboard tubes and a marble along with graph paper and a criteria checklist. Four computers, connected to the internet were available for the students to use. Clarification of task requirements was provided to each group by a mathematics educator, however no direct teaching of task content was provided.

9.4.3. *Data analysis*

Bakhtin's (1986) notion of 'voice' was used to analyse the transcripts. Bakhtin (1986) formulated a theory of voice that emphasised the active, situated, and functional nature of speech as it is employed by various communities within a particular society. Taking the notion of 'utterance' rather than 'word meaning' as a basic unit of communication, Bakhtin maintained that in dialogue with others, people align themselves within different speaking positions or voice types as they produce or respond to an utterance or a chain of utterances. Such voice types reflect the social ways of communicating that characterise various group behaviours (for example, professional communities, age groups, and socio-political authorities) that a person has had the opportunity and/or willingness to access. As such, 'voice' as used here, encompasses "what" is being said, the "way" in which it is spoken, and the positioning of speakers in relation to the authority framework established within the communication.

9.5. Analysis and Discussion

We enter the mathematics modelling challenge when Helen, Nicole, Neil and Aaron (a student from a different school to the others) are preparing a short journal entry of their experience with the challenge. The extract is

Table 9.1. *Maybe our whole group learnt about it.*

Turn	Speaker	Text
01	Nicole	Aaron learnt... What did you learn?
02	Aaron	I learnt lots.
03	Nicole	Well then tell us.
04	Aaron	I learnt about slope.
05	Neil	Maybe our whole group learnt about it.
06	Helen	I didn't (learn about slope), I had to do it ($y = mx + 3$).
07	Neil	(I learnt) About the equations.
08	Helen	$y = mx + 3$

taken from the second targeted data collection session held in the morning of the second day of the challenge (see Table 9.1).

We enter the script where Nicole is recognising Aaron's 'belonging' in the group by asking him what he had learnt from engaging with the mini-golf task. Instead of accepting Aaron's general response (Turn 2 – *I learnt lots.*) and then moving to record the responses of the other members of the group who were from her school, Nicole encourages Aaron to be reflective and to consider the specifics of what he had learnt (Turn 3 – *Well then tell us.*). This action reflects a reason for using the practices of collective argumentation in the modelling challenge, "We want to encourage our students to be reflective and consider how the various concepts (in mathematics) are related".

Collective argumentation benefits this level of understanding by requiring students to explain and justify their learning on a regular basis, therefore, making knowledge public. Explaining and justifying involve the gathering and sharing of evidence that satisfy disciplinary constraints associated with coherence and logic. Explaining and justifying allow students to become conscious of others' ideas and points of view, allowing processes of thought as well as products to become visible.

This privilege of reflecting on understanding continues as the other members of the group comment on their personal understandings relating to 'slope'. Here we see students being reflective, considering what they have learnt (Turn 7 – *About the equations.*) and what they did not learn (Turn 6 – *I didn't, I had to do it.*). However when students say they have learnt if, it does not mean they understand it, as Helen reveals in the next sequence of text (see Table 9.2).

Table 9.2. *I knew something that you didn't know.*

Turn	Speaker	Text
09	Nicole	Helen, what did you learn today?
10	Helen	$y = mx + 3$
11	Nicole	Didn't you already know that?
12	Helen	No, how to do it, like I knew what it (slope) was, I just didn't know how to do it (slope).
13	Nicole	Didn't you know how to do it (slope)?
14	Helen	You (Nicole) didn't.
15	Nicole	Yes I did, well I knew how to do it the obvious way, I knew how to do it on a graph, but on quadrant things (quadrants of a full grid).
16	Neil	I knew something that you didn't know.

Here we see Helen and Nicole linking what they know, considering a different strategy (using $y = mx + 3$ [Turn 10] or graphing a line on a grid [Turn 15]) and recognising they are doing the same thing. Through this text, we see Helen and Nicole selecting and transferring the mathematical tools they had learnt in the classroom to this context and recognising that there are different ways of applying those tools and different levels of knowing about and using mathematical tools.

This form of teaching and learning privileges the recognition of multiple representations of a mathematical idea through requiring students to individually represent a solution or idea about a task and to compare their representation with others. When students complete a brief written response to a text, or a solution to a problem, or an evaluation of the effectiveness of an experiment, they are more likely to participate in any discussion that follows, ask questions of others, share ideas with others, and to self-monitor their understanding (Gaskins, Satlow, Hyson, Ostertag & Six, 1994). Comparing representations allows students to see what is the same and what is different about their ideas and interpretations. In the process, it can help students learn by making them view concepts from different perspectives, and can be affirmed when students see congruence between ideas and representations (Feltovich, Spiro, Coulson & Feltovich, 1996).

Through recognising that Nicole is using a graphical approach to completing the task and that Helen is using an algebraic approach, the students

Table 9.3. *We used it, but we didn't know how.*

Turn	Speaker	Text
17	Nicole	Neil, what did you learn?
18	Neil	I learnt that, I learnt just that ($y = mx + 3$).
19	Nicole	What do you mean just that? $y = mx + 3$?
20	Neil	Just write everybody learnt that ($y = mx + 3$), because we all did learn that, yeah everybody learnt it.
21	Nicole	I need an eraser.
22	Neil	So you don't have to write just Aaron (learnt $y = mx + 3$) cause we all learnt it.
23	Nicole	Did anyone else learn anything that's not there (in the journal entry)?
24	Neil	Um maybe we ...
25	Nicole	How to use slope or anything?
26	Helen	That (slope) is part of the equation.
27	Neil	Yeah, that's part of the equation. Let's see, what about how to ...
28	Aaron	Did you know that equation ($y = mx + 3$) before we came (to the challenge)?
29	Nicole	We used it ($y = mx + 3$), but we didn't know how.
30	Helen	That's how to find out 'm'.

pave the way for relating procedural to conceptual understanding, as illustrated in the following sequence (see Table 9.3).

Once again (Turns 17 & 19) a member of the group, Neil, is asked by Nicole to explicate what he learnt from engaging in the mini-golf task. Neil's admission that he learnt about slope (Turn 22 – *So you don't have to write just Aaron cause we all learnt it.*) marks a moment in the conversation when this group of students from two different schools, have become a group who are willing to take ownership of their learning. In so doing, links are made between *'how to use slope'* (Turn 25) and the algebraic equation $-y = mx + 3$ (Turn 26 – *That is part of the equation.*) and between the concept of 'slope' and its algebraic representation 'm' (Turn 30).

Collective argumentation benefits the linking of conceptual with procedural understanding and linking individual with collective understanding by requiring that the group attain consensus about a response to a task that they can present to the whole class — a response that each member of the group understands. Consensus based on understanding is the end product of a process of considering and critiquing. Students negotiating a common

Table 9.4. *Let me explain how to do 'm'.*

Turn	Speaker	Text
31	Nicole	Yeah, I got that, I got it ($y = mx + 3$).
32	Neil	Yeah I do (understand).
33	Nicole	Because I didn't really get it ($y = mx + 3$) before, but I understand now.
34	Gail	So you (Helen) just explained how to do '*m*'?
35	Neil	Let me explain how to do '*m*'.
36	Gail	He (points to a boy in her group) needs to figure out also how to do '*b*' (the *y* intercept).
37	Helen	$mx + 3$, equals '*b*' equals *y* intercept.
38	Gail	He (a boy in her group) doesn't get (understand) it.
40	Helen	Whatever *Y* intercept is, is '*b*'.
41	Gail	He (a boy in her group) doesn't understand.
42	Helen	*Y* intercept is when the *Y*, where the point *Y* is. Well then you (Neil) can explain it then.

understanding of a representation or idea take learning from the co-operative to the collaborative plane of learning.

This willingness to collaborate in the sharing of understandings continues in the next sequence of text which was recorded in a moment of interest when the group extended its boundaries to include members from other groups undertaking the challenge. We enter the script where Helen has just explained to her group how to find the slope '*m*' in the equation $y = mx + 3$ (see Table 9.4). During the explanation, students from other groups gathered around to listen. The students included Gail (another student from Nicole, Helen and Neil's school).

The interaction of students in the above text is interesting for students engaged in an inter-school mathematics challenge. The challenge relating to the mini-golf course can be won by one group only. Helen, in demonstrating her understanding of how to find the slope of a line between two points provides an explanation that is attended to by students not in her group. Not only does Helen share her understanding with Nicole (Turn 31) and Neil (Turn 32), but also receives a request from Gail (a member of another group) to explain again how to find the slope, as a boy in Gail's group does not understand. Neil requests permission to provide the explanation (Turn 14). However, Helen simply revoices the main point of her explanation (Turn 16). Upon receiving a signal from Gail that this revoicing is insufficient (Turn 19), Helen gives Neil permission to explain. Neil goes on to present an explanation to the gathered audience that results in a number of students from different groups working together to build a model that they can use to make predictions.

Collective argumentation benefits a view of mathematics of being an engagement in communal practice by requiring each group to present their agreed approach to the class for discussion and validation. Such presentations of group work permit students to engage with the conceptual content of a lesson at their level, to employ their own prior experiences, preconceptions and language, and to distribute the nature of their knowing across a group rather than in a fashion that focuses on any one individual.

9.6. Conclusion

This chapter sets out to describe how mathematical modelling can provide for students a view of mathematics where the mathematics they use has relevance outside the classroom. While building a mathematical model, students must do much more than practise a set of procedures. They need to interpret the context, they need to identify suitable assumptions, using the mathematics they need to look for consistency in the mathematics and consensus between the mathematics and the context. If they do not find consistency and consensus, they then need to revisit the context and consider why their model is not a suitable representation. Once the students have the model they are then interested to know whether the model can be used to make predictions.

As teachers involved in the mathematical modelling we need to know what the students are thinking. Further, as well as developing an understanding of the mathematics we need to ensure students are able to operate together to build on each other's understandings reaching a level of sophistication very few students would naturally achieve on their own.

The nature of the learning displayed by Helen, Nicole and Neil as they engaged with the mathematics challenge in designing a hole-in-one mini-golf course, suggests that these students view doing the mathematics as providing a forum where personal understandings can be expressed, re-considered, shared and co-authored — an unusual stance for students engaged in what might be viewed as a mathematics competition. The above analysis of student-student interaction suggests that within this group's way of doing the mathematics challenge, understanding emerged around shared practice. That is, a collaborative space emerged where a voice of inquiry enacted that privileged: (a) the relating of conceptual understanding to procedural understanding (e.g., determining slope and the y intercept to build the equation, $y = mx + b$), (b) group ownership of learning over individual performance (e.g., designating new learning about slope to the whole group rather than just to an individual within the group), (c) mathematising, that is, not only knowing the mathematics, but also how and when to use the mathematics (e.g., as suggested by Nicole's statement — *We used it* ($y = mx + 3$) *but we didn't know how*).

Within this collaborative space, on-going processes for adding meaning to the mini-golf task such as representing, comparing and explaining were used by the students in a fashion that allowed their individual representations, ideas and points of view to become products of the moment, able to be used by others to progress understanding. Students' interactions, as portrayed in the above conversations, imply that within the collaborative space constructed by the students within the constraints of the mathematics modelling challenge, students not only co-constructed knowledge, but developed also an awareness of the 'self' as operating with tools of mathematics (e.g., $y = mx + b$), of the self operating as a mathematician.

The interactions between Helen, Nicole, and Neil occurred within a real novel context centred on real mathematics challenges. Rather than displaying individual personalities engaged in competitive intellectual practice, Helen, Nicole and Neil were drawn into a culture of inquiry that displayed distinct co-operative and collaborative relationships. This culture is further testified through student feedback forms about the modelling challenge collected over the past five years that have attested to students enjoyment to a group based modelling process. In response to the question, "What did you find most enjoyable about the mathematical modelling challenge?", students consistently referred to working with their peers and engaging with challenging mathematical tasks.

This chapter has provided a sense that teachers can change students' views of the learning mathematics through engaging students in mathematical modelling. Such engagement can not only lead to students talking about and doing mathematics outside the classroom, but also provide teachers with the means to change their sense of self-efficacy in the classroom. Supported by practices such as those associated with collective argumentation, teachers may utilise tasks that not only permit students to generalise and objectify their thinking, employ practices that promote consistency in their reasoning and consensus in their ways thinking, but also create contexts where students can safely discuss their ideas, accept guidance from others, and reflect upon their own practice.

References

American Mathematical Society: http://www.ams.org/profession/career-info/math-work/math-work, 30.11.11.

Bakhtin, M. M. (1986). *Speech Genres and Other Late Essays.* (C. Emerson & M. Holquist, Eds., V. W. McGee, Trans.). Austin: University of Texas Press.

Boaler, J. & Greeno, J. G. (2000). Identity, agency and knowing in mathematics worlds. In J. Boaler (Ed.), *Multiple Perspectives on Mathematics Teaching and Learning* (pp. 171–200). Westport, CT: Ablex.

Brown, R. A. J. & Renshaw, P. D. (2000). Collective argumentation: A sociocultural approach to reframing classroom teaching and learning. In H. Cowie and G. van der Aalsvoort (Eds.), *Social Interaction in Learning and Instruction: The Meaning of Discourse for the Construction of Knowledge* (pp. 52–66). Amsterdam: Pergamon Press.

Feltovich, P. J., Spiro, R. J., Coulson, R. L. & Feltovich, J. (1996). Collaboration within and among minds: Mastering complexity, individually and in groups. In T. Koschmann (Ed.), *CSCL: Theory and Practice of an Emerging Paradigm* (pp. 25–44). Mahwah, N. J.: Lawrence Erlbaum Associates.

Galbraith P. L., Stillman G. & Brown J., Turning Ideas into Modelling Problems, In Lesh R., Galbraith P. L., Haines C. R. & Hurford A. (Ed.), *Modelling Students' Mathematical Modelling Competencies*, ICTMA13, 133–144, Springer.

Galbraith, P. (1989). From applications to modelling. In D. Blane & M. Evans (Eds.), Mathematical Modelling for the Senior Years (pp. 78–86). Parkville: The Mathematical Association of Victoria.

Galbraith, P. (1995). Assessment in mathematics: Developments, innovations and challenges. In L. Grimison & J. Pegg (Eds.), (1995). *Teaching Secondary School Mathematics.* (pp. 289–314). Sydney: Harcourt Brace.

Gaskins, I. W., Satlow, E., Hyson, D., Ostertag, J. & Six, L. (1994). Classroom talk about text: Learning in science class. *Journal of Reading*, 37(7), 558–565.

Ginsburg, H. P. & Seo K-H. (1999). Mathematics in children's thinking. *Mathematical Thinking and Learning*, 1(2), 113–129.

Lingefjard, T. (2006). Faces of Mathematical Modelling. Zentralblatt Fur Didaktik der Mathematik, 38(2), 96–112.

National Council of Teachers of Mathematics (NCTM). (2000). *Principles and standards for school mathematics*. Reston, VA: Author. Retrieved from http://www.standards.nctm.org.

Pajares, F., & Graham, L. (1999). Self-Efficacy, Motivation Constructs, and Mathematics: Performance of Entering Middle School Students. *Contemporary Educational Psychology*, 24, 124–139.

Perkins, D. N. (1992). *Smart Schools: Better Thinking and Learning for Every Child*. New York: The Free Press.

Smith, J. P. III (1996). Efficacy and teaching mathematics by telling: A challenge for reform. *Journal for Research in Mathematics Education*, 27(4), pp. 387–402.

Swetz, F. & Hartzler, J. S. (1991). *Mathematical modelling in the secondary school curriculum*. The National Council of Teachers of Mathematics: Reston, Virginia.

Van Oers, B. (1998). From context to contextualizing. *Learning and Instruction*. 8(6), pp. 473–488.

Vygotsky, L. (1987). Thinking and Speech. In R. W. Rieber & A. S. Carton (Eds.), *The collected works of L. S. Vygotsky, Volume1: Problems of general psychology*. New York: Plenum Press.

Wertsch, J. V. (2002). *Voices of collective remembering*. New York: Cambridge University Press.

Wilhelm, J., Baker, T. & Dude, J. (2001). *Scaffolding adapted from strategic reading: Guiding students to lifelong literacy*. New Hampshire USA.: Heinemann, a division of Reed Elsevier Inc., <http://www.myread.org/scaffolding.htm>.

Section 3

Mathematical Learning through Modelling Tasks

CHAPTER 10

LEARNING THROUGH "DESIGNING A CAFE"

YEO Kai Kow Joseph

Over the last six years in Singapore, the mathematics curriculum was revised to place emphasis on reasoning, communications and connections; applications and modelling in addition to heuristics and thinking skills as processes that encompass the implementation of the mathematics curriculum. Its implication is that even secondary school students should now be involved in works of mathematical modelling. This chapter discusses an attempt by a group of Secondary 2 students who were undergoing a four-day modelling outreach to model a floor plan of a cafe. The three emergent themes arising from the case study were: (1) the experience of creative and cooperative problem solving, (2) the learning experiences of teachers and students through mathematical modelling, and (3) the mathematical modelling as problem solving from a modelling perspective.

Keywords: Secondary School, Mathematisation, Modelling Process, Measurement, Geometry.

10.1. Introduction

In mathematics education, for more than two decades, problem solving has been looked upon as the fundamental goal of mathematical instruction (Schoenfeld, 1994). Since the 1980s, problem solving in the school mathematics movement has been gaining momentum around the world. The 1980 Yearbook of the National Council of Teachers of Mathematics (Krulik, 1980), indicated that problem solving would be the theme of the 1980s. Also in 1980, the importance of problem solving in mathematics teaching and learning was highlighted by the publication of the Agenda for Action (NCTM, 1980). One of its recommendations was that "problem solving must be the focus of school mathematics" (p. 2). In Australia, these views were written in A National Statement on Mathematics for Australian School (Australian Education Council, 1990), where problem solving was one of four sub-headings in the Mathematical Inquiry Strand. In New Zealand, the importance of problem solving has been emphasised by the Ministry of Education's (1995) publication, Implementing Mathematical Processes in Mathematics in the New Zealand Curriculum. The open-ended approach (Becker & Shimada, 1997), which is a widely practised instructional method in Japanese mathematics classes, centres on problem solving. Moving in the same direction, Singapore's Ministry of Education revised the syllabus for the mathematics curriculum in schools with a primary aim to enable students to develop their abilities in mathematical problem solving (Ministry of Education, 2006). In Singapore the primary purpose of teaching mathematics is to enable students to solve problems. This aim is dependent on five factors; specifically, skills, concepts, attitudes, metacognition and processes. The new framework of the Singapore Curriculum continues to encompass mathematical problem solving as its central focus. There are some changes in the inter-related components, such as "reasoning, communication and connections" and "applications and modelling" being part of processes (Ministry of Education, 2006, p. 13), a position, which is in line with similar 21st reform-based visions of schooling around the world (NCTM, 2000; NSW Board of Studies, 2002). The deliberate inclusion of the new mathematical processes in Singapore is seen as a significant step towards making problem solving more relevant. The Singapore's mathematics curriculum states that "mathematical modelling is the process of formulating and improving a mathematical model to represent and solve real-world problems and that students should learn to use a variety of representations of data, and to select and apply appropriate mathematical methods and tools in solving real-world problems" (Ministry of Education, 2006, p. 14). These new inclusions are by no means a minor extension to the curriculum. To promote mathematical processes such as these will require alternative pedagogies and assessments that sharply contrast with conventional ones.

10.2. Mathematical Modelling as Problem-Solving Activity

A modelling perspective to mathematical problem solving focuses on the students' representational fluency through the flexible use of mathematical ideas where the students have to make mathematical descriptions of the problem situation and data. When students paraphrase, explain, draw diagrams, categorise, find relationships, dimensionalise, quantify, or make predictions, they are generally developing their conceptual systems or models through the mathematising. As they work with the rich contextual data, they would need to surface and communicate their mathematical ideas to clarify their thoughts and weigh the validity of their ideas. During the modelling process, a student interprets a given real-world problem within his or her socio-cultural context and form his or her own mental image of the problem. Next, various possible perspectives at representing the problem mathematically focusing on factors affecting the outcome of the problem and the relationships between the selected factors and the outcome of the problem are considered. In this chapter, the following instructional design principles from the works of Lesh, Cramer, Doerr, Post and Zawojewski (2003) are used as criteria as well as in the designing of mathematical modelling task:

1. The reality principle — the modelling task should resemble some everyday life experiences and require sense-making and broaden prior knowledge.
2. The model construction principle — the modelling task expects students to test, modify or encompass a mathematically key construct.
3. The self-evaluation principle — the modelling task requires students to self-evaluate.
4. The construct documentation principle — the modelling task expects students to manifest their thinking about the task.
5. The simple prototype principle — the model is sufficiently easy to utilise.
6. The model generalisation principle — the result functions as a worthwhile model for interpreting other similar settings.

Many perspectives of mathematical modelling abound. The modelling task used for the purposes shared in this chapter adheres to the "contextual modelling perspective" (Kaiser & Sriraman, 2006, p. 306) where decision making during the modelling process factors in real-world experiences and integration of socially constructed knowledge across disciplines. This perspective of modelling is chosen because it is of direct relevance to the current focuses of the Singapore mathematics curriculum: the activation of the processes of reasoning, mathematical communication, and connections during problem solving. In particular, connections refer to linkages within mathematical topics, between mathematics and other subject-disciplines, and between mathematics and everyday life experiences.

The purpose of this chapter is to discuss mathematical modelling as problem solving and models of one group of Secondary 2 students in their engagement with a model-eliciting task. Some implications and challenges are also discussed with respect to implementing modelling activities in the Singapore mathematics curriculum.

10.3. A Case Study: Modelling a Café Problem

This case study reports a group of Secondary 2 students (13 to 14 years old) working on a mathematical modelling problem as part of their modelling activity during a four-day mathematical modelling outreach where the author was the task supervisor. The modelling task was completed over a period of three days. Mindful that this was the first time that all the three Secondary 2 students were engaged in mathematical modelling, the task supervisor provided guidance through discussions until the completion of the modelling task. Prior inputs about the mathematical modelling process and problem solving were included in the modelling outreach.

10.3.1. *Modelling task: A café problem*

The café problem shown in Figure 10.1 adequately meets the design principles criteria presented earlier.

1. Real-world situation (Reality) — The context takes the form of a situation where the students actually have to work as a team to design and

One of the classrooms in the school is to be converted into a café. The purpose of the café is to encourage interaction among members of the school.
Your team is invited to:
(1) design and draw the floor plan for the café, and
(2) propose how the café can be furnished within the given budget of $30 000.
The floor plan should be drawn to scale and labelled with the various parts of the cafe and their actual real-life dimensions. Furnishings for the café include furniture items and appliances.

Figure 10.1. The café problem.

draw the floor plan of a café. The café problem was contextualised to make the problem-solving process take place in a natural setting. The students were in the actual classroom and they were supplied with 1-cm square papers, scissors, tapes, markers, rulers and calculators and they had to make a floor plan that was drawn to scale.

2. *Identifying the variables* (model construction) — The students explored the problem by discussing the meanings of 'furnishings' and 'designing' and looking at ways that a café can promote interaction among its customers through its layout and furnishings. The students initially identified the following ways that a café can promote interaction: (1) the distance between the customers, (2) the arrangement of furniture, (3) the ease in moving around in the café, (4) the dimensions of the tables and chairs, (5) the design of the tables and chairs, and (6) the width of the walk way between the furniture.
3. *Clarifying assumptions* (model construction and self-evaluation) — The students made the following assumptions: (1) the whole classroom floor area could be fully utilised, (2) the café is for students, (3) the installation and delivery cost were included in the cost of furniture.
4. *Forming relationships* (model construction) — The students determined their own measurements of the classroom layout and made a relationship between the floor area and size of furniture. They also worked on the cost of the furniture within the given budget of $30 000.
5. Testing and revising the model (self evaluation) — The students drew sketches of different arrangement of the furniture and calculate the cost to narrow their range of the arrangement and cost to attain a café that encourage interaction.
6. *Showing the model* (construction documentation) — The students designed a sketch of a café to show how it may look like. The students also explained how the arrangement of the furniture may affect interaction among customers.
7. Transferring the knowledge (generalisation) — The students drew to scale a floor plan and labelled the various parts of the cafe and their actual real-life dimensions. Other considerations included the space between each set of furniture as well as the cost, type and dimensions of furniture used. This is to mathematically achieve their goal instead of aimlessly sketching different floor plan (see Figure 10.2).

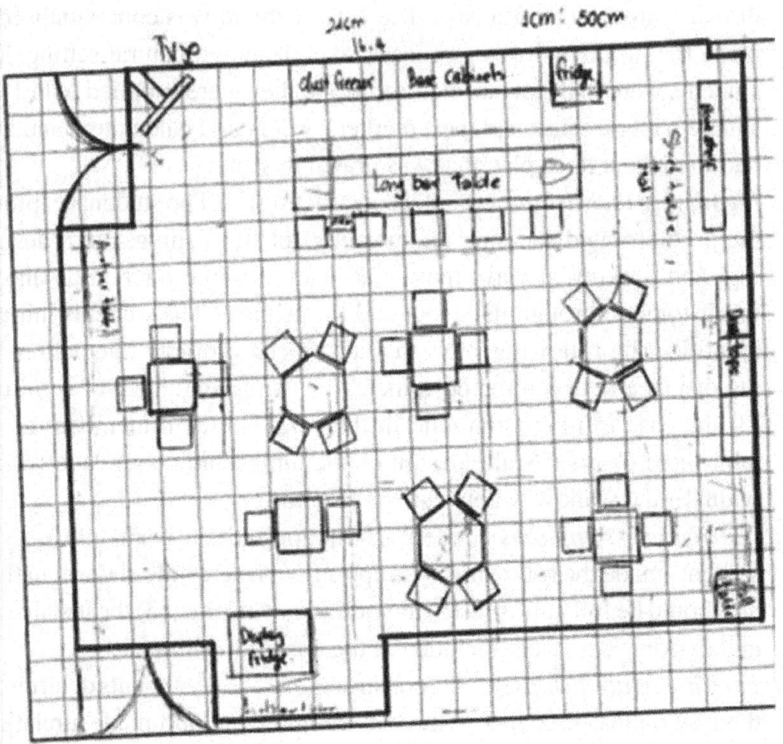

Figure 10.2. Scale drawing of café.

In the café modelling task, the students pondered on mathematical knowledge related to geometry and measurement. They had made sketches for the floor plan and also realised that there are many different floor plans that could meet the criteria. They had engaged in conjecturing about the type and cost of furniture that would give a better design of café. They had also related how the dimensions of the tables and chairs would affect the floor area of the café and work towards plausible dimension combinations to designing a café that encourage interaction. In the solution shown in Figure 10.2, the students wanted to design a café such that had different combinations of seating arrangements. Relying on their understanding of shape of table, they devised a creative way of getting people to be seated in different ways. They have tables for 2 to 6 people to be seated around so as to encourage interaction. They are also mindful of the cost of each piece of furniture and worked within the budget of $30 000. The learning value here was not just in terms of the conceptualising the representation and making the mathematical translation but in comparing different sketches of floor plan and reasoning why their solution had achieved the objectives of the modelling task.

10.4. Pedagogical Implications

Three themes and issues emerged in the conduct of this case study. First, through mathematical modelling the students were able to experience the creative and cooperative nature of problem solving. Mathematical modelling had allowed the Secondary 2 students to experience the cooperative and creative nature of learning mathematics.Even though the Secondary 2 students indicated the mathematical modelling task challenging, they were able to complete it through teamwork and perseverance. The modelling task further allowed the students to learn about the creative nature of mathematics as they worked on the mathematical model which allowed their thinking to move from the real-world situation to the mathematical world. The Secondary 2 students were able to make mathematical abstractions from a real-world situation. It was exciting to note the misconceptions and misconnections that emerged as the students grappled with their mathematical ideas and equally promising to see how they resolved the demanding areas among themselves. The misconceptions and misconnections that were surfaced should be seen as opportunities for refining of initial ideas. As observed in the Café Problem, students were discussing the various scale drawings and chose the best floorplan to meet the goals of the modelling task. Moreover, their explorations within the model construction stage gradually led them to make critical conjectures and generalisations. Strategies such as drawing different sketches were evident as the students intended to optimise the floor area of the café during their explorations.

Secondly, when introducing mathematical modelling in schools, student learning becomes a very significant consideration. Since there is always a certain amount of uncertainty when we allow students to find creative solutions in doing modelling, it is thus essential for teachers themselves to experience the entire process of modelling. The experience thus gained would give the teachers the confidence to guide the students in their modelling task. It is imperative that the teacher is able to make sound judgements on when to provide scaffolding, what scaffolding to provide, and how to provide scaffolding. In addition, teachers should create instructional activities using modelling tasks that will not only enable students to internalise the mathematical processes but also promote increased student proficiency in mathematics. The case study presented here may assist mathematics teachers to create classroom environments that foster mathematical reasoning and connections.

The third emergent theme of this case study is mathematical modelling, as problem solving from a modelling perspective does not insist that students get the correct answer. In given contexts or problem situations, there are various specific questions that could be posed in a modelling task. For each of these, different assumptions and definitions will unavoidably result in multiple 'correct' solutions. In this Café Problem and many others, for example, what is implied by the purpose of the café is to encourage interaction among members of the school and this is best defined in this particular context. In mathematical modelling, the context underpins everything and is the purpose for engaging mathematically. However, this is not necessarily the case in problem solving in general. It has been acknowledged that students' initial conceptions developed during the modelling process could be characterised by unwarranted assumptions or the imposing of inappropriate constraints (English & Watters, 2004) and that their reasoning could also be seen as uncoordinated and simple (Lesh & Doerr, 2000). The problems they run into should not be viewed as being falling into a trap but as opportunities for refining their preliminary results. As observed in the Café Problem, the "flawed" reasoning that appeared was not reproached but rather valued because it exposed the students' misconceptions and it provided opportunities for others to discuss those ideas. What was crucial was that through the modelling process the students were able to take control of what they believed would lead to their model solution. As compared to closed problem solving, the modelling perspective emphasised process rather than product.

10.5. Conclusion

Lower secondary school students need rich opportunities to develop their mathematical reasoning and connections abilities; modelling tasks provide such opportunities. To implement modelling activities in the secondary mathematics classroom, it is needful for teachers to have an understanding of what it is, how it is different from conventional pedagogies, and what the learning outcomes look like. There are, however, numerous challenges if modelling activity is to become a reality in the classroom. Undoubtedly, it will require highly competent teachers who have mastered the complexity of the lesson where the students are continuously being actively engaged in constructing and applying the mathematical ideas and skills. Teachers' skills and knowledge are important to consider as well as the level of support and resources from the school administration. One main challenge therefore is to develop teachers' skills, knowledge and attitude to implement modelling activities in the lower secondary mathematics classroom. While formal training through workshops and seminars may be able to impart new knowledge to teachers, more discussions between teachers and mathematics educators would be required to bridge the gap between theory and practice. Teachers need support to acquire the skills and confidence to implement modelling activities at the classroom level.

References

Australian Education Council. (1990). *A National Statement on Mathematics for Australian Schools*. Australia, Melbourne: Curriculum Corporation for Australian Education Council.

Becker, J. P. & Shimada, S. (1997). *The Open-Ended Approach: A New Proposal for Teaching Mathematics*. Reston, VA: National Council of Teachers of Mathematics.

English, L. D. & Watters, J. J. (2004). Mathematical modelling with young children. In M. J. Hoines & A. B Fuglestad (Eds.), *Proceedings of the* 28th *Annual Conference of the International Group for the Psychology of Mathematics Education* (Vol. 2, pp. 335–342). Bergen, Norway: PME.

Kaiser, G. & Sriraman, B. (2006). A global survey of international perspectives on modelling in mathematics education. ZDM — *The International Journal of Mathematics Education*, 38(3), 302–310.

Krulik, S. (1980). *Problem Solving in School Mathematics: Yearbook*. Reston, VA: National Council of Teachers of Mathematics.

Lesh, R., Cramer, K., Doerr, H. M., Post, T. & Zawojewski, J. S. (2003). Model development sequences. In R. Lesh & H. M. Doerr (Eds.), *Beyond Constructivism: Models and Modelling Perspectives on Mathematics Problem Solving, Learning and Teaching* (pp. 35–58). Mahwah, NJ: Lawrence Erlbaum Associates.

Lesh, R. & Doerr, H. M. (2000). Symbolizing, communicating, and mathematizing: Key components of models and modelling. In P. Cobb, E. Yackel & K. McCain (Eds.), *Symbolizing and Communication in Mathematics Classrooms: Perspectives on Discourse, Tools, and Instructional Design* (pp. 381–284). Mahwah, NJ: Lawrence Erlbaum Associates.

Ministry of Education. (1995). *Implementing Mathematical Process in Mathematics in the New Zealand Curriculum.* Wellington: Learning Media.

Ministry of Education. (2006). *Secondary Mathematics Syllabus.* Singapore: Curriculum Planning and Development Division.

National Council of Teachers of Mathematics (NTCM). (1980). *An Agenda for Action: Recommendations for School Mathematics of the 1980's.* Reston, VA: National Council of Teachers of Mathematics.

National Council of Teachers of Mathematics (NCTM). (2000). *The Principle and Standards for School Mathematics*, Reston, VA: NCTM.

NSW Board of Studies NSW (2002). *Mathematics K-6 Syllabus 2002.* Sydney, New South Wales, Australia: Board of Studies, NSW.

Schoenfeld, A. H. (1994). Reflections on doing and teaching mathematics. In A. H. Schoenfeld (Ed.), *Mathematical Thinking and Problem Solving* (pp. 53–70). Hillsdale, NJ: Erlbaum.

CHAPTER 11

LEARNING THROUGH "PLANE PUNCTUALITY"

HO Weng Kin

The prompt service of in- and out-bound flights has made Changi Airport one of the best airports in the world. This chapter showcases how modelling tasks can be woven into the rich fabric of real-life contexts, that is familiar to Singapore students.

Keywords: Secondary School, Facilitation Stages during Modelling, Queueing Theory.

11.1. Overview

Practitioners in applied mathematics would agree, to a large extent, that a mathematical model is a description of a system using mathematics as the vehicular language. The process of manufacturing and developing a mathematical model is termed mathematical modelling. Here, *modelling* denotes unambiguously mathematical modelling. The use of mathematical models is ubiquitous, ranging from the natural sciences to the social sciences.

Because modelling yields 'usable' representations of any existing system (Eykhoff, 1974), it is natural to ask whether it can offer a versatile platform for mathematics learners to apply a wide-ranging repertoire of mathematical skills. Pertaining to the integration of modelling into school mathematics curricula, the aforementioned question has been addressed in recent works such as Stillman *et al.* (2007), English and Watters (2004) and English (2004). These works advocate that the "modelling process is driven by the desire to obtain a mathematically productive outcome for a problem with a genuine real-world motivation" (Galbraith & Stillman, 2006, left of p. 143). By engaging the students in the process of modelling, the designed activities are intended to "motivate, develop and illustrate the relevance of particular mathematical content" (Galbraith, Stillman, & Brown, 2006). In this chapter, we shall adopt this perspective of modelling.

11.2. The Singapore Scene

The importance of using "mathematical modelling as content" (Julie, 2002, top of p. 3) in Singapore mathematics education has been continuously emphasised in schools since its inception in 2003. The Ministry of Education's official formulation of the 4-stage cycle: (1) Mathematisation, (2) Working with mathematics, (3) Interpretation and (4) Reflection (Balakrishnan, Yen & Goh, 2010) echoes this emphasis. Subordinate to this cycle, the task designer identifies an interesting context upon which the modelling task can be constructed (Kaiser & Sriraman, 2006). Whence, the choice of a Singapore example is a necessary one: Singapore school students must first be able to readily identify with national icons, second to work through the problem, and lastly to appreciate the real-life applicability of textbook mathematics. Note that this idea has been exploited in (Wong, 2003) as a possible source of reinforcement in National Education under the label of "Homeland", albeit in somewhat the opposite direction.

11.3. This Study

The present study seeks to identify some key factors that contribute directly towards the success of mathematical modelling activities, focusing on both the process and the product, in the Singapore secondary school context. We shall explain what we mean by 'success' in our ensuing discussion.

11.4. The Method

Five groups of Secondary 2 (equivalent to Year 8 of the Australian or US education system) students were assigned a common modelling task 'Plane Punctuality' (see Figure 11.1), and they were to carry out the task over three days (1–3 June 2010). Amongst these were two mixed groups consisting of students belonging to different schools. On 3 June 2010, all groups were to present (for about 10–15 minutes) their research findings and recommendations to their peers. In those three days, two facilitators

The Singapore Changi Airport (SIN) is the main airport in Singapore and a major aviation hub in South-east Asia. SIN is currently ranked among the best 19 airports in the world. How the punctuality of arrivals and departures is managed is an important task the airport must undertake every day. How *on-time* are flights in SIN compared to other well-known international airports?

Figure 11.1. The modelling task "Plane Punctuality at Singapore Changi Airport".

(who had received *a priori* a one-day training in facilitation of mathematical modelling activities) facilitated and monitored the students' progress. These facilitators provided minimal direct assistance.

The task requirement was intentionally open-ended. For instance, students were free to interpret what 'on-time' meant. Also, details such as the types of flights, time-frames of observation, and the explicit list of 'other well-known international airports' were not provided. The task design was based on the 5 + 1 design principles developed in Galbraith (2006). In particular, by applying the Didactical Principle to craft the guiding questions one derives these labels: (a) Discussion, (b) Plan, (c) Experimentation (Data organisation), (d) Representation and verification and (e) Product. These design principles guide the development of mathematisation skills, communication skills, reasoning abilities, critical thinking, skills in data representation and organisation. To a large degree, students were free to venture in their courses of investigation. This characterises the sort of learning that takes place in a modelling setting, i.e., one which focuses on the direction (reality $\rightarrow$ mathematics), where one tries to locate the appropriate piece(s) of mathematics to help one solve a real-life problem (Stillman *et al.*, 2007).

With the aim of this study in mind, field observations hoped to capture the behaviour, learning outcomes, modelling processes and final product of two groups, **A** and **B**. Each group had three members. **A**'s members came from the same school, while **B** was a mixed group. A *successful* modelling experience would ideally be one in which the *meaningful application* of the aforementioned modelling *processes* results in the developing of *non-trivial insights* into the real-life situation one is trying to model. The field study is to identify those factors which directly contribute to such a success.

11.5. Students' Mathematical Modelling Experience

11.5.1. *Discussion*

Flight punctuality, defined by **A**, was "being on time or with a maximum of 15–30 minutes delay". **B** defined flight punctuality as flights being able to arrive or depart an airport within the desired period of time, where 'desired period of time' was qualified to be 15 minutes from the schedule timing. **B** allowed the possibility of a plane arriving or departing earlier than scheduled, and also identified punctuality as a concept tied to frequency. Hence **B** demanded punctuality to entail a minimum of 90% of the total flights meeting the above requirement.

The groups' goals were different. While **A** set out to maximise punctuality in terms of ground operations, **B** sought for a comparison between Singapore and other international airports. The 'discussion' aspect of the task design provided evidence of the groups' attempts to understand and simplify the problem, e.g., **B** decided to carry out their study focusing only on arrivals. Decision-making like this constantly occurred in the entire modelling endeavour. Both the groups identified the variables affecting the chosen goal. **A** singled out crew effectiveness and airport layout, while **B** looked at location of the peak periods in a day. The process of identifying the relevant variables affecting the chosen goal moved the students towards mathematising the problem they were trying to solve. This was then followed by formulation of underlying assumptions, such as "the passengers are cooperative" and "there were no terrorists" for **A**, and "the weather is fine", "the (political) condition of the country is stable" for **B**. Seino (2005) advocates that the awareness of assumptions "plays the role of a bridge that connects the real world and the mathematical world" (Seino, 2005, top of p. 665) and serves as an effective teaching principle. **A** articulated such awareness in their log-book: *"(in) modelling ... the answers are neither* *wrong nor right, quite a lot of things we need to assume it, and when the answer is out we still need to check whether it is reasonable for us, but we like challenging."*

11.5.2. *Plan*

The next phase was planning. Question 7 in the handout facilitated this change of phase by probing "How does your team plan to do their investigations about the task?"

Changes made in plans subsequently took place for both groups: **B** moved the step of identifying the peak period of air traffic in a day from (3) to (1), while **A** made a much sharper refinement (See Figure 11.2).

Part One: Discussion

7. How does your team plan to do your *investigations* about the task?
Include key steps to be taken and how work is to be distributed among the members.

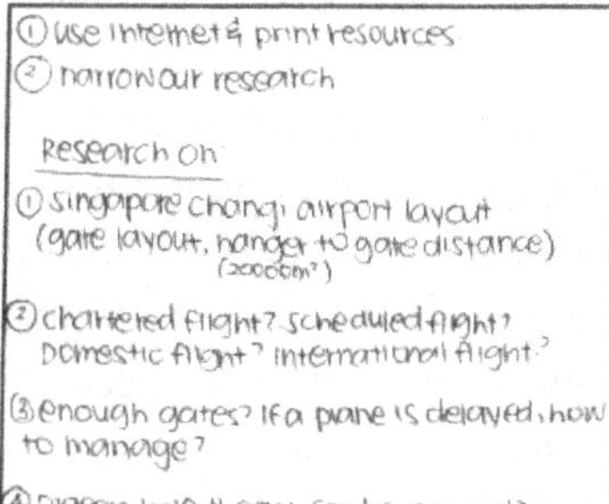

① use internet & print resources
② narrow our research

Research on
① singapore changi airport layout (gate layout, hanger to gate distance) (20000m²)
② chartered flight? scheduled flight? Domestic flight? international flight?
③ enough gates? If a plane is delayed, how to manage?
④ pigeon hole theory. can be applied?

Part One: Discussion

7. How does your team plan to do your *investigations* about the task?
Include key steps to be taken and how work is to be distributed among the members.

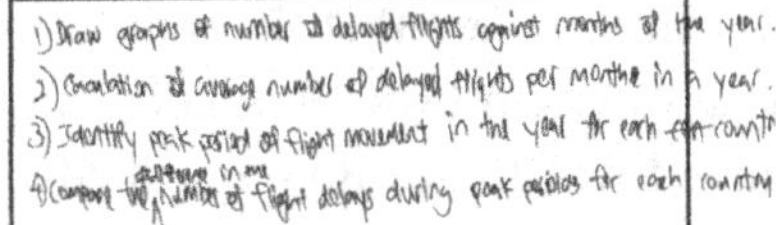

1) Draw graphs of number of delayed flights against months of the year.
2) Calculation of average number of delayed flights per months in a year.
3) Identify peak period of flight movement in the year for each country.
4) Compare the average number of flight delays during peak periods for each country.

S/No.	Items	Remarks
1.	Research on reasons why planes may have been delayed	• No place for planes to park • Ineffective crew (not guiding the planes where and how to land) • Drawing graphs of number of delayed flights is not enough.
2.	Efficiency of other terminals	• Put all statistics into a table format • Just the average number of delayed flights is insufficient for statistical inferences
3.	Consequences of planes being late	
4.	How to solve problems	• Speed flow density relationship • Queuing theory, Pigeonhole theory

Figure 11.2. Group A's plan refinement (top left: before 2nd meeting; bottom: after 2nd meeting); Group B's plan (top right: before 2nd meeting)

The action of 'fine-tuning' and 'adjusting' their plans were evidence of the non-linearity and cyclical nature of the key modelling activities (Doerr, 1997). Keywords used by the students, such as "graphs", "number of delayed flights", "difference", "speed flow density relationship", "statistics", "queuing theory" and "pigeonhole theory" were also evidence of the students' attempt to mathematise.

11.5.3. *Experimentation, data organisation, representation and verification*

Experimentation, data organisation, representation and verification observed in **A** and **B** became very much intertwined and cyclical. Here, one focuses on the *diversity* of the outcomes evolved.

Group A. First, this group set out to verify that SIN was indeed ranked among the top in terms of its flight punctuality. From a reliable internet source, they obtained the punctuality rates of the top 5 major international airports, and compared these with SIN (See Table 11.1):

An easy computation of the average yields 75.82%, thus confirming the claim. Aware that the maximum data point is always higher than the mean (hence making this calculation an overkill), what the students really wanted was to see how much SIN deviated above the average. With the aim to find out the possible causes of flight delay, **A** then deduced from the available data that about 5000 arrivals and departures in SIN occurred every week, i.e., yielding an average of 714 flights per day. **A** then collected data from Flight statistics (n. d.) on 1 June 2010, reporting 375 departures and 369 arrivals. They wrote: *"Out of these flights, 50% of the departing flights were punctual, while 88% of the arriving flights were on time. This totals up to 159 delays in departure and 41 delays in arrivals, and gives us an average of 15 flights each for both departure and arrival terminals per hour."*

Table 11.1. *Punctuality rates of top 5 international airports in 2010.*

Country	Singapore	Tokyo	Rome	UK	India
Puntuality Rate	92%	90%	83.1%	65%	49%

Then **A** deduced that "30 gates will be in use" in each hour. Then crucially, they realised that while being parked at a gateway, planes required some turnaround time which involved "disembarking the passengers, unloading and reloading baggage, re-fuelling and cleaning the plane before passengers could board again and set off to their destination".

Delays, as **A** perceived, were caused by the violation of a simple rule: at anytime, the number of available gates must be at least equal to the number of incoming planes that need the gates. **A** recognised this to be an instance of the *pigeonhole principle*. Motivated by this principle, **A** worked out (using a layout diagram of the 92 gateways in SIN) the maximum turnaround time for just enough gateways available to park/service the planes. Since the airport is populated by 30 planes in each hour, it would have taken about 3 hours to have all the gateways occupied if one assumes that all the airplanes which were parked at the gateways had not left yet. By track of reasoning, **A** deduced that the turnaround time could at most be 2 hours.

Group B. **B** went on to identify the peak periods for arrival for 5 airports: (1) Hong Kong International, (2) Suvaranabhumi International, (3) Changi International Airport, (4) Kuala Lumpur International Airport and (5) Beijing Capital Airport. In addition, **B** compared the data from SIN with that from the Hong Kong International Airport: "The 5.7% of airplane delays at the SIN could be caused by unexpected weather conditions and also the timing of departure from its origin country. The low 2.2% of delayed flights at HK Airport might also be due to luck and good weather conditions on that day".

11.5.4. *Product*

Group A. **A** extended their scheduling argument to the seating schedule for the passengers. Guided by queuing theory, **A** proposed to seat the passengers efficiently and quickly by first seating the people situated furthest from the doors, and the people in the window seats, before proceeding 'outwards'. Two seating arrangements were proposed (see Figure 11.3). Also, baggage due for transit should be placed near the door of the cargo compartment.

Group B. For their product, **B** gave qualitative suggestions:

1. Airport personnel should be properly trained.

Back of plane

1	8	15
2	9	16
3	10	17
4	11	18
5	12	19
6	13	20
7	14	21

Front of plane

Back of plane

1	2	3
4	5	6
7	8	9
10	11	12
13	14	15
16	17	18
19	20	21

Front of plane

Figure 11.3. Group A: Two proposed seating arrangements based on queuing theory.

2. Reduce waiting time for passengers.
3. Increase the airport's customer service satisfactory level.

11.6. Findings and Implications

The keen reader must have noticed by now the presence of a significantly wide gap in the sophistication of modelling tools used by **A** as compared to **B**. A probe into **A**'s background knowledge revealed that they were mathematically gifted students trained for the Singapore Mathematics Olympiad. Also, prior to this activity, one particular member of **A** had completed a mathematics project (where she used queuing theory to model traffic congestion near her school). In this respect, **B** lacked advanced mathematical training/exposure.

It is clear that of the two groups, **A** had done a better job. While it would be easier to account for **A**'s success by simply appealing to their larger mathematical expertise, one should perhaps relook at the aspect of *mathematical modelling competency*. Here, mathematical modelling competency refers to one's ability to identify relevant questions, variables, relations or assumptions in a given real-world situation, to translate these into mathematics and to interpret and validate the solution of the resulting mathematical problem in relation to the given situation, as well as the ability to analyse or compare given models by investigating the assumptions being made, checking properties and scope of a given model (Niss, Blum, & Galbraith, 2007).

A made it a habit to check the properties and scope of a chosen model from time to time, while the same was not observed of **B**. For example, having decided queuing theory as a mathematical model, **A** looked for more data, i.e., the data of the timing between the passengers' loading and unloading, to confirm the appropriateness of such a model. Recorded in their daily log-book was this: *"First, we need to obtain more data for more accurate analysis. Next we can cater to a specific peak period where there are more airplanes arriving and departing. Also we should cater directly to the different airplane sizes so specific gates can be allocated to different models of planes, and thus, the appropriate equipments and technicians would be on hand."*

Mathematical modelling is a versatile approach to meaningful learning of mathematics, where modelling tasks and questions can be readily crafted around day-to-day experience in any given culture and country. The flight punctuality problem described above serves as one such exemplar. Though localised to a small sample of 6 students, our study seems to indicate that mathematical modelling competency is probably the key factor that determines the diversity in behaviours and learning outcomes derived from a single mathematical modelling activity. Thus, a classroom teacher who wishes to exploit mathematical modelling should certainly bear in mind the current modelling competencies of the students he or she is teaching.

References

Balakrishnan, G., Yen, Y. P. & Goh, E. L. E. (2010). Mathematical applications and modelling. In B. Kaur & J. Dindyal (Eds.), (pp. 247–257). World Scientific.

Doerr, H. M. (1997). Experiment, simulation and analysis: an integrated instructional approach to the concept of force. *International Journal of Science Education*, 19, 265–282.

English, L. (2004). Mathematical modeling in the primary school. In I. Putt, R. Faragher & M. McLean (Eds.), Proceedings of the 27th annual conference of MERGA (pp. 207–214).

English, L. & Watters, J. (2004). Mathematical modeling in the early school years. *Mathematics Education Research Journal*, 16(3), 59–80.

Eykhoff, P. (1974). *System Identification: Parameter and State Estimation*. Wiley & Sons.

Flight statistics. (n.d.). Retrieved from http://www.flightstats.com.

Galbraith, P. (2006). Real world problems: developing principles of design. In P. Grootenboer, R. Zevenbergen & M. Chinnappan (Eds.), Proceedings of the 29th annual conference of the MERGA (Vol. 1, pp. 229–236).

Galbraith, P. & Stillman, G. (2006). A framework for identifying student blockages during transitions in the modelling process. *ZDM — The International Journal of Mathematics Education*, 38(2), 143–162.

Galbraith, P., Stillman, G. & Brown, J. (2006). Identifying key transition activities for enhanced engagement in mathematical modelling. In P. Grootenboer, R. Zevenbergen, & M. Chinnappan (Eds.), Proceedings of the 29th annual conference of the MERGA (Vol. 1, pp. 237–245).

Julie, C. (2002). Making relevance relevant in mathematics teacher education. In I. Vakalis, D. H. Hallet, D. Quinney & C. Kourouniotis (Eds.), Proceedings of the 2nd International Conference on the Teaching of Mathematics: [ICTM-2]. New York: Wiley. [CD-ROM]

Kaiser, G. & Sriraman, B. (2006). A global survey of international perspectives on modelling in mathematics education. *ZDM — The International Journal of Mathematics Education*, 38(3), 302–310.

Niss, M., Blum, W. & Galbraith, P. (2007). How to replace the word problems. In W. Blum, P. Galbraith, H.-W. Henn & M. Niss (Eds.), Modelling and Applications in Mathematics Education: The 14th ICMI Study (pp. 3–32). New York: Springer.

Seino, T. (2005). Understanding the role of assumptions in mathematical modeling: analysis of lessons with emphasis on 'the awareness of assumptions'. In P. Clarkson *et al.* (Eds.), Proceedings of the 28th annual conference of the MERGA (pp. 664–671).

Stillman, G., Galbraith, P., Brown, J. & Edwards, I. (2007). A framework for success in implementing mathematical modelling in the secondary classroom. In J. Watson & K. Beswick (Eds.), Proceedings of the 30th annual conference of the MERGA (pp. 688–697).

Wong, K. Y. (2003). Mathematics-based National Education: A framework for instruction. In S. K. S. Tan & C. B. Goh (Eds.), Securing our Future: Sourcebook for Infusing National Education into Primary School Curriculum (pp. 117–130). Singapore: Pearson Education Asia.

CHAPTER 12

LEARNING THROUGH "MOBILE PHONE PLAN"

CHENG Lu Pien

CHUA Kwee Gek

Keywords: Secondary School, Data Handling, Teacher Education.

12.1. Introduction

Problem solving is the heart of the Singapore Mathematics curriculum framework. Teachers are encouraged to engage students in various mathematical problems, including those situated in real world contexts. Mathematical modelling tasks are perceived to be important within the curriculum framework as they provide platforms for students to sharpen their mathematical reasoning and communication processes. At the same time, these tasks highlight connections between school mathematics and the real world [OECD, 2012]. In this chapter, the notion of modelling reflects the views of Lesh and Zawojewski [2007] and Zawojewski [2010] on problem solving and modelling; that is, modelling involves a process of interpreting a problem-solving situation where "a system of interest needs to be represented by a mathematical system — which will simplify some things, delete others, maintain some features, and distort other aspects" [p. 238]. In other words, modellers are to make appropriate assumptions, set parameters and select variables to work within the given context of the real-world problem based on their interpretations of the situation at hand.

A mathematical model is referred to as a "mathematical representation or idealisation of a real-world situation. It can be as complicated as a system of equations or as simple as a geometrical figure" [CPDD, 2012a]. What is important is that the model should share as many features as the real-world situation as it seeks to represent and propose a solution. In the process of developing a mathematical model as reported in this study, modellers (students aged 13–14) embark on a cyclical modelling process as they learn to formulate, solve, interpret and reflect on their models within the premises of the real-world problem. Educators view this cyclical process which engages students in mathematisation of the real-world as a crucial aspect of mathematical literacy [see de Lange, 2006].

According to PISA 2012 draft framework, *mathematical literacy* is an individual's capacity to formulate, employ, and interpret mathematics in a variety of contexts [OECD document November 2012]. de Lange [2006, p. 17] describes mathematisation as key to mathematical literacy. This study adopted the mathematisation cycle by de Lange [2006, p. 17]. To explain the modelling process, four components are identified: real-world problem, mathematical problem, mathematical solution and real-world solution. The modelling process starts with a problem situated in reality. The problem solver first tries to understand the problem, makes assumptions to simplify the problem and sets up a model based on reality. Next, the problem solver tries to solve the mathematical questions within the mathematical model. Lastly, the problem solver interprets the mathematical results obtained in the context of the real world and validates the solution. This study adopts a Model and Modelling Perspective (MMP) [see Lesh & Doerr, 2003] in task design, implementation and facilitation. The MMP perceives mathematical models as "purposeful conceptual systems that are expressed using some (and usually several) representational media, and their purposes generally are to describe or to explain the behaviours of other systems so that intelligent predictions, constructions, or manipulations can be made" [Lesh & Doerr, 2003, p. 44]. In this respect, MMP sees modellers making symbolic representations of a real-world situation during modelling development as opposed to the traditional mathematical learning where they make meaning out of given symbolic representations. The MMP advocates the use of model-eliciting activities. "Model-eliciting activity is defined as a problem-solving activity constructed using specific principles of instructional design in which students make sense of meaningful situations, invent, extend and refine their own mathematical constructs" [Kaiser and Sriraman, 2006, p. 306]. When participating in a model-eliciting activity, modellers engage in processes such as "quantifying, dimensionalising, coordinating, organising, systematising or (in general) mathematising objects" [Lesh,

2003, p. 44]. This study intends to conduct a two-pronged investigation into the implementation of a model-eliciting task during the Mathematical Modelling Outreach event [see Chapter 1 of this book]: to (a) study the ***mathematical thinking*** involved or surfaced by the student-modellers working in groups in the development of their respective models and (b) identify ***key focuses in teacher education for facilitation of students' modelling competencies***.

12.2. Facilitating Students' Modelling Competencies

Kaiser and Grünewald [in this book, Chapter 2] define modelling competencies as "the ability and the willingness to work out problems with mathematical means taken from the real-world problem through mathematical modelling". As articulated by Maaß [2006], the modelling process is closely aligned with modelling competencies. The complexities of the modelling process prompted Maaß to present a multi-dimensional definition of modelling competencies. Aspects of the competencies include understanding the real-world problem, developing a model based on the problem, mathematising the model, solving the mathematical problem, interpreting the mathematical results in a real situation and validating the solution. Research on how teachers in Singapore facilitate students' modelling competencies is limited because schools are only beginning to focus on modelling tasks at secondary levels in recent years. This study took place during the Modelling Outreach Event held in June 2010 [in this book, Chapter 1]. A six-stage facilitation structure [see Ng, 2011] was implemented for this research. In this study, we are interested in the key focuses related to facilitating students' competencies in mathematical thinking and mathematisation of the real-world problem during modelling. This chapter hopes to address the nature of students' mathematical thinking during model development in the *Phone Plan Task* and identify some key focuses in teacher education for the facilitation of students' modelling competencies.

12.3. Significance of Study and Method

This chapter reports our observations of two pre-service undergraduate teachers' actual facilitation of the mathematical modelling process with a group of Secondary 2 students on the *Phone Plan Task*. Samples of the students' work and the mathematical thinking observed were also reported. We hope that by examining samples of the students' work and discussing the challenges faced by the facilitators in their facilitation journey will provide teachers who are new to facilitation of mathematical modelling some insights on how to leverage on the elements that facilitate the growth of mathematical modelling process through problems arising from real-world situation using real-world data.

The *Phone Plan Task* shown below is a problem situated in reality. Deciding on which phone plan works best for the different groups based on their usage experiences and budget is an important aspect of financial literacy. Understanding the *Phone Plan Task* requires the problem solver to make certain assumptions — identify the key variables and quantify other compounding variables that influence the ***best*** plan for the various groups. To begin the task, the problem solver has to source for information such as various phone plans offered by the different telecommunication companies. They should be able to sieve out relevant information within the allocated time frame; for example, the three main categories of consumers — teenagers, working professional and housewives. We define the categories of consumers as our key variables.

Phone Plan Task

In Singapore, having a mobile phone is deemed as a necessity for communication. It is a handy and trendy tool. Telecommunication companies in Singapore offer attractive packages catering to all walks of life throughout the year. Propose a **best** *"toolkit" guided by the following groups of consumers: teenagers, working professionals and housewives to decide on what is best for them.*

An important criterion in choosing a phone plan is budget (one relevant quantity) and the benefits offered by each plan. When choosing a plan, one must be clear of the kind of services that one uses more often than the others e.g., short message service (another quantity). The information provided for various phone plans by the different telecommunication companies is far too complex for the students to choose the appropriate mathematical notations or representations to model the information. Hence, setting up a mathematical model requires the problem solver to simplify the relevant quantities and establish their relationships so as to reduce the complexity of the situation. After drawing out and simplifying the relevant quantities, the problem solver proceeds to gather live data from the different groups of consumers to substantiate their models and solutions. Based on the collated information, the problem solver proposed a ***best*** toolkit for the different users and communicated their findings through an oral and poster presentation.

The Phone Plan Task was assigned to 15 Secondary Two students in five groups of three. Two pre-service undergraduate teacher facilitators were assigned to facilitate the modelling processes. Prior to the event, they attended an induction programme conducted by the advisors to familiarise them with the mathematical modelling processes. During the actual implementation, the advisors were mainly observers writing field notes of the facilitators' attempts in facilitating the task, intervening their facilitation when they sensed that they could not cope with queries from the groups. Data sources to examine the nature of mathematical thinking during the model development process of the Phone Plan Task included all the students' artefacts (written) and field notes by the advisors. The questionnaire completed by the facilitators at the end of the event was used to identify some key focuses in teacher education for the facilitation of students' modelling competencies.

12.4. Findings

12.4.1. *Nature of students' mathematical thinking*

We classify mathematical thinking manifested by the students' to the *Phone Plan Task* into two categories; namely, students' collection and representation of data and students' views about mathematical thinking.

12.4.1.1. Students' collection and representation of data

All the groups googled for the telecommunication companies in Singapore and decided to use the same three main telecommunication companies for the problem. Four common quantities were identified — local SMS, local incoming and outgoing calls, internet and data bundle. They then carried out surveys to collect most current data on the usage volumes from the three consumer groups. Survey participants were mostly the students' relatives, teachers, school mates or friends. The surveys varied from printed survey forms, text messages from handphones or chat messages from the computers. Based on the data collected and services provided by the telecommunication companies, the five groups produced the ***best*** toolkit. The ***best*** toolkit consisted of recommended plans for the different consumer groups. They ranged from simple table form, ICT via online blogs [Figure 12.1], to visual display of data via line graphs [Figure 12.2].

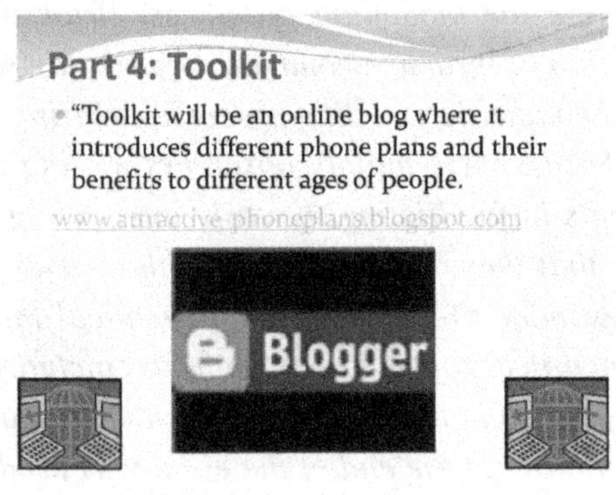

Figure 12.1. Online blog.

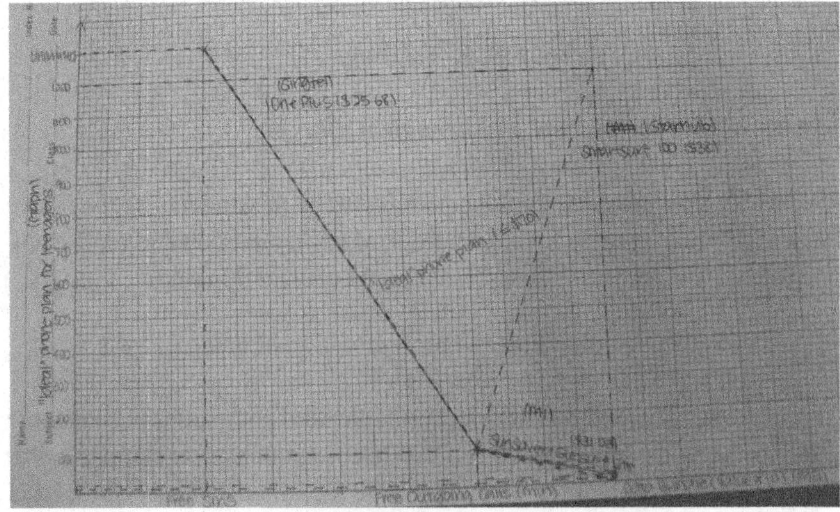

Figure 12.2. Line graph.

The various representations (tables, graph and on-line blogs) generated by the students reflect the students' mathematical thinking as they try to make sense of the real-world data.

12.4.1.2. Students' views about mathematical thinking

Our observations of the students' engagement on the modelling processes are that the open-endedness of the task (especially this is their first attempt) makes it difficult for students to fully enact the mathematisation process. Most of the students had difficulties "seeing" the mathematics in the task which, did not directly involve the use of formula or algorithms. The *Phone Plan Task* differed from the standard tasks found in the Singapore secondary school textbooks. As a result, two groups "had difficulty formulating [and picking up] the problem situation" [Schaap, Vos and Goedhart, 2011, p. 143] because they were unable to "find relevant directions and cues in the problem text to help them proceed with the task" [p. 143]. They lack certain modelling competencies and this is reflected in their work on the *Phone Plan Task*. They are clueless as to how to begin with the given task. They expected more hints, guidelines and more data in the problem to help them recognise quantities and identify the variables. Four groups were stuck after identifying the variables because the task does not have any numerical numbers. The students did not anticipate that mathematical problem solving and thinking include the need to source for their own data and filter the relevant data from the vast amount of information collected.

12.4.2. *Key focuses in teacher education for the facilitation of students' modelling competencies*

Critical scaffolds were necessary to help two groups identify and gain deeper understanding of the three categories of consumers. Findings reported below are implications related to pedagogies for facilitation of open-ended tasks and group work.

Pre-service teachers may have certain mind sets related to what mathematics is, which may impede the implementation and facilitation of open-ended tasks like a modelling task. For example, the current pre-service teachers may have a view that mathematics is rigid and prescriptive and that the mathematics solution process is linear, amounting to one and only one correct answer. However, the *Phone Plan Task* is divergent in nature. As such, the facilitators need to recognise and address multiple interpretations and varied solutions arising from the task. Facilitator B wrote that "in a mathematical modelling task, students [set] their own goals. Since the goals set by everyone may vary, this means that the solutions provided by them will be different too" [questionnaire].

Pre-service teachers need to be equipped with questioning skills targeting on getting students to articulate their reasoning, and mathematical thinking. They need to learn how to harness students' initial interpretations to the real-world problem and help move students on in their preferred pathways of investigations. The *Phone Plan Task* has great potential for multiple interpretations and depending on the students' entry levels; it is capable of generating different levels of questions. This uncertain dynamics in the facilitation process proves to be very challenging.

Not only do the pre-service teachers need to manage the progress of the different groups, they also need to know how to move between episodes of facilitating whole class discussions and group discussions. In particular, flexibility in the facilitation process needs to be exercised to ensure that all the students benefit from the mathematical modelling task. For example, switching from the guidance stance to '*explicit teaching*' for those who have problems understanding and managing the task. Facilitator B wrote, "it will be a challenge to conduct mathematical modelling task in real classroom situations [due to varied ability of students and the large number]".

12.5. Discussion and Conclusion

In this study, we examined key focuses related to facilitating students' competencies in mathematical thinking and mathematisation of the real-world problem during modelling. The results show that the mind sets of teachers about mathematical thinking needs to change as a first step towards building such competencies. The pre-service undergraduate teacher facilitators took on new and unfamiliar roles in interacting with the students through the *Phone Plan Task*. They learnt that the facilitation of modelling tasks requires them to learn, listen and observe the students before asking 'appropriate' questions that would further students' understanding and clarification of the problem situation. This result is in accordance with Doerr's [2006] findings which suggest that to engage students in the development of mathematical models, the mind sets of teachers need to be changed such that they are more ready to listen to students' thinking and respond in ways that enable students to further develop their emerging models. Results from this study also show that viewing mathematical thinking as inquiry and problem finding is necessary before embarking on the *Phone Plan Task.* Some degree of intuitive knowledge and understanding of mathematical concepts inherent in the task are also required of the students before they could proceed with the modelling task. Without these, the students would not be able to find the problem in the *Phone Plan Task.* Neither would *they* be able to recognise the *mathematics* involved.

The facilitation process can be very challenging in the modelling process due to varied interpretations of the task. The student facilitators not only have to answer queries related to the task but must also approach a fluid teaching stance when addressing the queries. Like Ng [2011], we observed that the coverage of mathematics by each group during the tasks depends on many factors. One factor that we observed, common to Ng, was task scaffolding. In particular, scaffolding in areas like, making appropriate assumptions for the problem situation and simplifying the problem is crucial to proceed with the modelling task. Similar to Ng, we also question how teachers can achieve a 'balance' in scaffolding such task and still retain the mathematical rigour. We also observe from this study that questioning skills is essential in the modelling process to harness students' initial interpretations to the real-world problem, stimulate students' development of metacognitive and critical reasoning skills, and promote effective mathematical communication.

References

Curriculum Planning and Development Division [CPDD]. (2012a). *Mathematical modelling resource kit.* Singapore: Ministry of Education.

de Lange, J. (2006). Mathematical literacy for living from OECD-PISA perspective. *Tsukuba J. of Educ. Study in Math.*, 25(1), pp. 13–35.

Doerr, H. M. (2006). Teachers' ways of listening and responding to students' emerging mathematical models. *ZDM — The Int. J. of Math. Educ.* 38(3), pp. 255–268.

Kaiser, G. & Grünewald, S. (in press). Promotion of mathematical modelling competencies in the context of modelling projects. In N. H. Lee & K. E. D. Ng (Eds.), *Mathematical Modelling: From Theory to Practice*. Singapore: World Scientific.

Kaiser, G. & Sriraman, B. (2006). A global survey of international perspectives on modelling in mathematics education. *ZDM — The Int. J. of Math. Educ.*, 38(3), pp. 302–310.

Lesh, R. & Doerr, H. M. (2003). *Beyond Constructivism: Models and Modelling Perspectives on Mathematics Problem Solving, Learning, and Teaching*. Mahwah, NJ: Lawrence Erlbaum Associates.

Lesh, R. & Zawojewski, J. S. (2007). Problem solving and modeling. In F. K. Lester (Ed.), *The Second Handbook of Research on Mathematics Teaching and Learning* (pp. 763–804). Charlotte, NC: Information Age Publishing.

Maaß, K. (2006). What are modelling competencies? *ZDM — The Int. J. of Math. Educ.*, 38(2), pp. 113–142.

Ng, K. E. D. (2011). *Facilitation and Scaffolding: Symposium on Teacher Professional Development on Mathematical Modelling — Initial perspectives from Singapore*. Paper presented at the Connecting to practice — Teaching practice and the practice of applied mathematicians: The 15th International Conference on the Teaching of Mathematical Modelling and Applications, Australian Catholic University, Melbourne, Australia. Note: this reference may be replaced with another which is the actual book chapter from the conference but this information may come at a later stage.

Organisation for Economic Co-Operation and Development [OECD]. (2012, 30 November 2010). Draft PISA 2012 mathematics framework Retrieved 2 Feb, 2012, from http://www.oecd.org/document/53/0,3746,en_32252351_32236130_46963701_1_1_1_1,00.html.

Schaap, S., Vos, P. & Goedhart, M. (2011). Students overcoming blockages while building a mathematical model: Exploring a framework. In G. Kaiser *et al.* (eds), Trends in Teaching and Learning of Mathematical Modelling, International Perspectives on the Teaching and Learning of Mathematical Modelling.

Zawojewski, J. S. (2010). Problem solving versus modelling. In R. Lesh, P. Galbraith, C. R. Haines & A. Hurford (Eds.), *Modelling Students' Mathematical Modeling Competencies: ICTMA 13* (pp. 237–244). New York: Springer.

CHAPTER 13

LEARNING THROUGH "THE BEST PAPER PLANE"

SOON Wan Mei Amanda

CHIOK Hwee Fen

KONG May Hua Maybelline

Local teachers receive little exposure to mathematical modelling (MM) and show apprehension towards this form of teaching. The aim of this chapter is to provide teachers with an illustration of what can transpire in a modelling activity, giving them a head start to MM. If teachers are well-informed about students' possible approaches to certain modelling activities, they can gain enough confidence to start conducting similar activities on their own. The task we focus here is the construction of a paper plane. Paper planes are instrumental in the process to moving on to progressively larger models of flight (for example, kites and aircrafts). What should the best paper plane be like? We report on the potentials and challenges during the modelling processes of students (aged 13–14 years old) undertaking the task.

Keywords: Secondary School, Graphing Calculator, Mathematical Reasoning, Model Development, Modelling Process.

13.1. Introduction

A mathematical modelling (MM) cycle consists of four elements, Mathematisation, Working with mathematics, Interpretation and Reflection (Balakrishnan *et al.* (2010)). It involves translating a real-world problem into a mathematical one by formulating a mathematical model. Students are required to make suitable assumptions, simplify the problem to a solvable one, identify the mathematical concepts and variables, set up a mathematical model such as drawing a graph function or a set of equations, select appropriate methods to solve the problem, interpret the solution and reflect on the model and mathematics that they used.

In the Annual Paper Airplane Contest (described in English & Watters, 2005), students have to design planes that win awards while meeting certain rules. One award is given to the team whose plane stays in air for the longest time; another to the team whose plane travels the greatest distance in a straight-line path; and the final award is given to the overall winner for the contest. Similarly, students undertaking the paper plane task depicted in this chapter needed to come up with planes that satisfy some rules. However, these students decide on their own notion of the "best plane", work through a modelling cycle, and the winners of the task are determined based on their overall performances during the entire modelling process.

Our task enables students to use an empirical approach in modelling (as discussed in Ang, 2001), where measurements are made through experiments and results are presented for example, in a graph. Students can engage in the process of generating quantities, operating on those quantities to provide a mathematical basis for decisions and reason the relationships among the quantities (Doerr & English, 2003). In addition, they can learn that the use of technology may simplify the execution of a task. Indeed, it can play an important role in the modelling process (Balakrishnan *et al.*, 2010).

However, as many students are new to modelling, they can feel helpless, insecure and lack orientation during the process (Kaiser & Schwarz, 2010). They may lack mathematical sense-making (English & Doerr, 2003) or possess a lack of understanding of mathematical concepts despite the capacity to correctly recall formulae (Stillman & Galbraith, 1998). Few students pay attention to verification of final solutions (Stillman & Galbraith, 1998). MM can easily be reduced to a mere curve-fitting exercise if students do not check that their solution obtained is valid in the real world (Galbraith, Stillman & Brown, 2006). Thus appropriate scaffolding provided by teachers can enable students to think more critically (Chan, 2008).

Singapore teachers are generally apprehensive in conducting modelling activities and need to be nurtured to accept various problem representations and diverse solutions (Ng, 2010). Getting acquainted with the problem task first would make it easier for a teacher to ease into the scaffolding role (Chan, 2008). It is our hope that this chapter can help teachers by providing them a picture of what transpired in a modelling task. In the following sections, we first describe the task at hand. Then we document the varied ways the students employed in doing the task, according to the potentials and challenges faced. Finally we discuss some implications and conclusions.

13.2. Paper Plane Task

The facilitators of this task were third year pre-service teachers and a faculty member, who implemented the facilitation structure proposed by Ng (Chapter 1 of this book). Five groups of 3 or 4 students (aged 13 to 14 years old) from different schools were asked to design the "best" paper plane within three days. As the majority of students were new to the modelling activity, the facilitators first engaged students through a simulation website http://flightsimx.archive.amnesia.com.au/.

Each group was then given one set of task sheets to record their discussion processes, sketches of planes they had folded, their experimentation processes, verification of how their plane is the "best" using mathematics, and reflections on the task. See Figure 13.1 for snapshots of some parts of the task sheets. The presentations were held on the last day of the event. Students were instructed to include the following points: *process of how they obtained their "best" plane; results to convince others why their planes were considered the "best"; highlight how they felt throughout the event, including the difficulties that they faced and how they were resolved.* A question-and-answer session then followed. Finally, the facilitators provided some feedback to the teams.

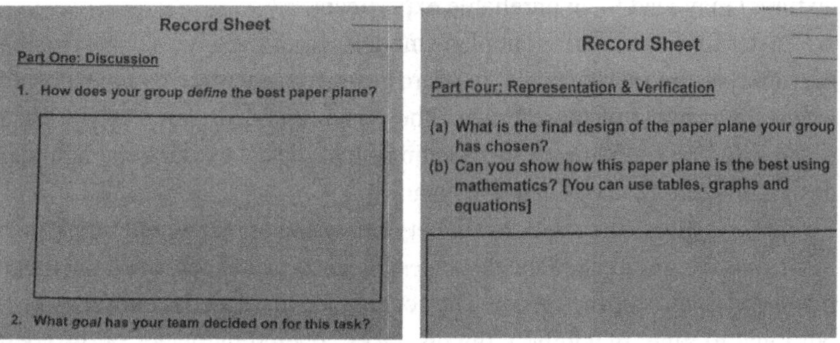

Record Sheet

Part One: Discussion

1. How does your group *define* the best paper plane?

2. What *goal* has your team decided on for this task?

Record Sheet

Part Four: Representation & Verification

(a) What is the final design of the paper plane your group has chosen?
(b) Can you show how this paper plane is the best using mathematics? [You can use tables, graphs and equations]

Figure 13.1. Snapshots of task sheets that students used.

13.3. Potentials of the Task

13.3.1. *Variables explored and relationships formed*

The variables considered by the teams included air resistance, shape, size and mass of plane, angle and force at which a plane is thrown, material of paper used, type of tip of plane (flat or pointed), length of plane's wings, location of throw (wind condition and height), number of folds of plane, design of plane, and throwing style of plane launcher. Most groups assumed same thrust force and angle launched, constant air resistance, and took an average of three trials for each experiment.

One group chose the "Hammer" design of plane (found in a guidebook on paper plane folding by [Stillinger, 2004]). They obtained a relationship between distance flown by plane and material of paper used. Materials considered were tracing paper, thin construction paper, thick construction paper and standard printing paper (all A4 sizes). It was concluded that a heavier plane results in greater stability of the flight (due to less obstruction from wind) and a speedier flight.

Another group investigated how the material of paper affects the total flight time in the air. They designed a plane which was not too long (for stability) and ensured the tip was not sharp (as they believed that a sharper tip causes greater speed and shorter time in the air). The materials were: construction paper, tracing paper, white printing paper, and corrugated paper. They obtained a quadratic relationship between mass of paper and the time flown. They also verified the relationship using another plane model. However they believed that the different materials (with different thickness and textures) may lead to an unreliable experiment.

After folding six different plane models (from [Stillinger, 2004]) using the same type and size of paper, the third group found that the hammer design led to the furthest distance flown. They then obtained planes of different wingspans and found the corresponding flight distances using the hammer design, to obtain the optimal wingspan.

To investigate the relationship between speed of planes and time flown in air, another group fixed the distances flown by planes, obtained the times flown and calculated the speeds. They concluded that speed is indirectly proportional to time. In addition, the distances covered by planes of different materials (corrugated, tracing and thick construction paper) were investigated. The A4-size corrugated paper was deemed the best.

The last group first folded 11 designs (from [Stillinger, 2004]) using construction papers of A4 size. They threw each plane twice, found the best two planes (Numbers 7 and 11) and tested them another time. Every member of the team then threw Plane 7 made using transparent, cardboard, corrugated and standard printing paper (normal paper). They looked for a common trend among the results when the plane was being thrown by different people. The trend was that corrugated paper was the best (using a normal throw), while the cardboard plane went the furthest in a "shot put style" throw. See Figures 13.3 and 13.4.

They explained that the plane made of corrugated paper was heavier and can propel itself through the air, and the average distance was "better" because the nozzle did not get damaged after each throw (since this paper was strong). Finally, a quadratic relationship between distance flown and mass of the plane helped the group to determine their best plane.

13.3.2. *Representations of data*

In considering the relationship between distance flown by plane and material of paper used, a table was created (Table 13.1) to give a clear representation of the data obtained.

A group who investigated how the material of paper affects the total flight time in the air plotted a graph of average time versus mass of plane (Figure 13.2).

Table 13.1. *Flight lengths versus different materials used.*

	Distance (cm)			
Materials Used	Plane 1 (tracing)	Plane 2 (thin construction)	Plane 3 (thick construction)	Plane 4 (printing)
Trial 1	786	680	822	565
Trial 2	817	709	855	490
Trial 3	745	650	785	591
Average	782.7	679.7	810.7	548.7

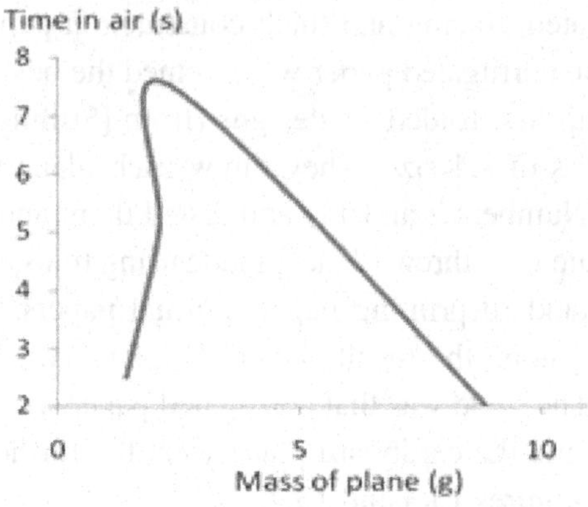

Figure 13.2. Flight time versus plane mass.

Simultaneous equations were solved manually to obtain a quadratic relationship between flight time (y) in seconds and mass of plane (x) in grams:

$$y = -0.136x^2 + 1.82x - 3.6.$$

From the graph, they concluded that there is an optimal mass of plane for the longest flight time.

The relationship between the wingspan of a plane and the distance flown was tabulated as given below:

Table 13.2. *Investigation of effect of wingspan on flight of paper plane.*

Wingspan (cm)	7.5	8.5	9.5	15.0	19.0	4.0	5.5
1st try (m)	9.1	8.2	7.5	7.0	7.3	7.6	11.2
2nd try (m)	9.0	8.2	7.5	8.0	7.1	7.7	10.6
3rd try (m)	9.8	7.9	8.1	7.8	7.2	7.8	10.3
Average (m)	9.3	8.1	7.7	7.6	7.2	7.7	10.7

Bar and line charts were used by the last team to showcase their results of the best plane design and best material of paper for different types of throwing styles of planes, as shown in the figures below.

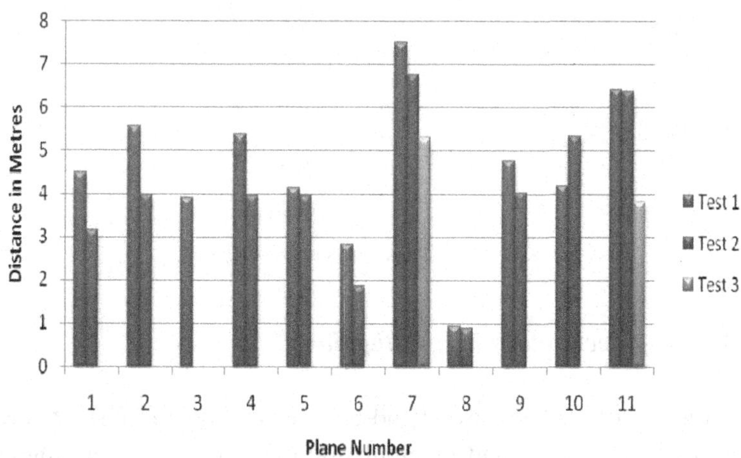

Figure 13.3. Distance vs plane types.

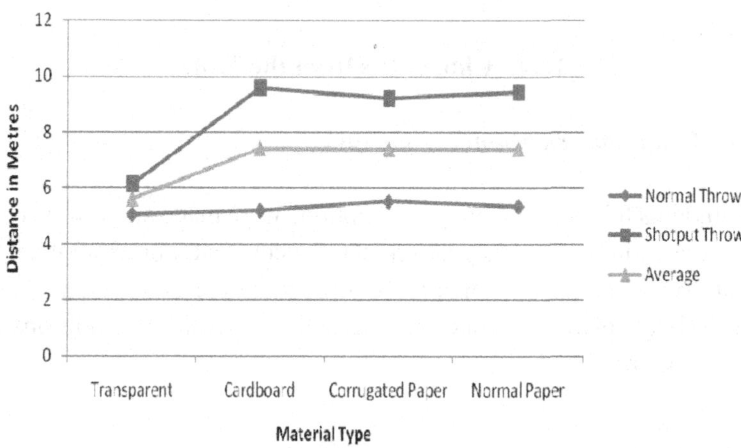

Figure 13.4. Distance vs throw types.

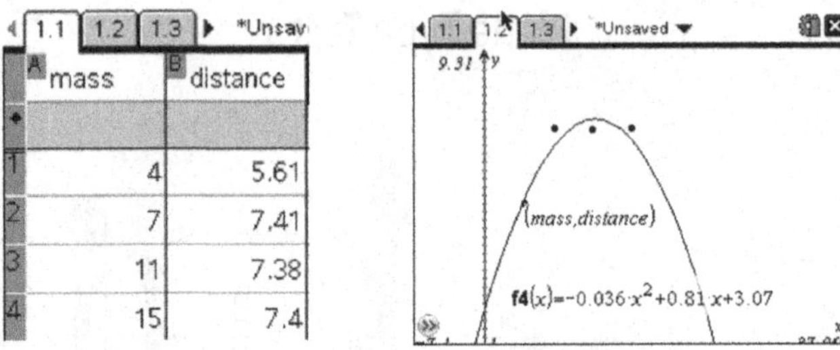

Figure 13.5. Distance vs mass of plane.

13.3.3. *Use of technology in investigations*

The last group of students used hand-held technology (TI-Nspire calculators) in their recordings and manipulation of data. It was convenient for recording information outside of the classroom when flying their planes. The quadratic relationship was also obtained much more easily as compared to other groups, using the "quadratic regression" function on their calculator (see Figure 13.5).

13.4. Challenges from the Task

13.4.1. *Conditions surrounding the task*

Most students felt that too many assumptions were made and it was hard to control the variables that they assumed to be constant. For example, it was difficult to use the same launch force throughout the experiment. Though they can fly the planes indoors, the air resistance and wind conditions may not be constant.

13.4.2. *Errors in use of data and deductions*

Though tracing paper is the lightest among all the materials tested, the distance travelled by the plane was not the shortest, as shown in Table 13.1. Thus the students failed to see that their conclusion of "heavier paper leads to longer distance flown" was not properly supported by their data.

From the relationship between flight time and mass of plane obtained, that is, $y = -0.136x^2 + 1.82x - 3.6$, we can see that zero mass implies a flight time of –3.6 seconds. Thus the model is inadequate. The students failed to evaluate the sense and meaning of this relationship within the real situation as they did not realise the importance of the domain of the function.

Similarly, the relationship between flight distance and mass of plane obtained, that is, $y = -0.036x^2 + 0.81x + 3.07$, was not subjected to the discussion of the domain of the function and the fact that zero mass implied a distance of 3.07 m (see Figure 13.5). In addition, they concluded that the best paper plane weighed 11g based on the graph obtained, instead of the actual numerical values. The process seemed to be reduced to a curve-fitting exercise.

The group who investigated the relationship between speed and time by fixing the distances flown by planes used the known formula (speed = distance/time) to calculate the speeds. So they concluded a known result while assuming the use of it.

The last group found that the cardboard plane went the furthest in a "shot put style" throw. They explained that it was because the material was strongest and the folds could stay (unlike the corrugated paper which was harder to fold, and as such came apart throughout the longer and more powerful flight). However, from a previous analysis, they found that corrugated paper was best for the normal throw as the material was the strongest. Hence there seems to be a discrepancy in their reasoning for the results.

13.4.3. *The modelling process*

During the event, we observed the following progress in the behaviour of students: *many did not know how to start*; *they had problems tabulating their results*; *some kept wondering if they were on the right track and thought of giving up on the task*; *many compared their task with their friends' and thought that other tasks were more meaningful*; *towards the end they seemed to appreciate the task and derived satisfaction from being able to figure out the task and complete it.*

We also noted that some of the groups did not work through the verification and generalisation processes, and appeared to be having difficulties linking the task to mathematics.

13.5. Implications and Conclusion

From this activity, several potentials were exhibited. For example, students were able to see the different variables affecting a task and understand the need for assumptions made. Some could draw relationships mathematically, make deductions and provide reasons. However there were also several challenges, including the failure of students to make sense of their solutions in the real world, weak reasoning skills, feelings of insecurities and helplessness. Thus we need to inculcate certain skills lacking in students through the MM processes.

As mentioned in Ang [2001], there is a need to rethink the approach to teaching some topics in mathematics, and Singapore teachers can attend in-service training to gain experience in MM. However, currently very few Singapore teachers have received any form of training for teaching MM in schools (Dindyal & Kaur, 2010). Teachers need to make sense of what MM processes mean and how they can be developed in students (Chan, 2008). Topics on MM should also be translated into teachable units for the teachers and incorporated in the regular curriculum (Dindyal & Kaur, 2010).

References

Ang, K. C. (2001). Teaching mathematical modelling in Singapore schools. *The Mathematics Educator*, 6(1), 62–74.

Balakrishnan, G., Yen, Y. P. & Goh, L. E. (2010). Mathematical modelling in the Singapore secondary school mathematics curriculum. In B. Kaur & J. Dindyal (Eds.), *Mathematical Applications and Modelling Yearbook 2010* (pp. 247–257). Singapore: World Scientific.

Chan, C. M. (2008). Using model-eliciting activities for primary mathematics classrooms. *The Mathematics Educator*, 11(1), 47–66.

Doerr, H. M. & English, L. D. (2003). A modeling perspective on students' mathematical reasoning about data. *Journal for Research in Mathematics Education*, 34(2), 110–136.

Dindyal, J. & Kaur, B. (2010). Mathematical applications and modelling: concluding comments. In B. Kaur & J. Dindyal, (Eds.), *Mathematical Applications and Modelling Yearbook 2010* (pp. 325–335). Singapore: World Scientific.

English, L. D. & Doerr, H. M. (2003). Perspective-taking in middle-school mathematical modelling: A teacher case study. In A. Neil Pateman, J. Barbara Dougherty & Joseph T. Zilliox (Eds.), *Proceedings of the 2003 Joint Meeting of PME and PMENA*, (vol. 2, pp. 357–364). Honolulu, HI: College of Education, University of Hawaii.

English, L. & Watters, J. (2005). Mathematical modelling in the early school years, *Mathematics Education Research Journal*, 16(3), 58–79.

Galbraith, P., Stillman, G. & Brown, J. (2006). Identifying key transition activities for enhanced engagement in mathematical modeling. In P. Grootenboer, R. Zevenbergen & M. Chinnappan (Eds.), *Identities Cultures and Learning Spaces* (Vol. 1). Canberra, ACT: Mathematical Education Research Group of Australasia.

Kaiser, G. & Schwarz, B. (2010). Authentic modelling problems in mathematics education — examples and experiences, *Journal for Didactics of Mathematics*, 31(1), 51–76.

Ng, K. E. (2010). Initial experiences of primary school teachers with mathematical modeling. In Kaur, B. & Dindyal, J. (Eds.), *Mathematical Applications and Modelling Yearbook 2010* (pp. 129–147). Singapore: World Scientific.

Stillinger, D. (2004). *The Klutz Book of Paper Airplanes*. Southam, England: Klutz Publishing.

Stillman, G. & Galbraith, P. (1998). Applying mathematics with real world connections: Metacognitive characteristics of secondary schools students, *Educational Studies in Mathematics*, 36, 157–189.

CHAPTER 14

LEARNING THROUGH "DESIGNING A TENT"

HO Siew Yin

This chapter reports a mathematical modelling task which was conducted during a 5-day Mathematical Modelling Outreach (MMO) event held in Singapore in June 2010. During the event, both primary and secondary school students in Singapore worked on various mathematical modelling activities. Students from Australia schools were also invited to take part in this mathematical modelling event. This chapter documents the mathematical modelling processes of the primary and secondary school students, as they worked on a modelling problem involving designing a 4-man tent for beach-goers. The mathematical processes were discussing, planning, experimenting and verifying.

Keywords: Primary School, Secondary School, Mathematical Thinking, Model Development, Modelling Process.

14.1. Introduction

In a world driven by globalisation and technological advancements, the workplaces have been transformed, and are still transforming into a place where there is an increase in demand for workers to possess competencies for the 21st century. Anecdotal evidence suggests that current employers are on the constant lookout for employees with 21st century skills (Jones, 2010; Lee, 2011). Examples of 21st century skills include civic literacy, global awareness and cross-cultural skills, critical and inventive thinking, information and communication skills (Ministry of Education, 2010). Hence employers expect their employees to, for example, be able to handle work "involving important mathematical processes . . . such as constructing, describing, explaining, predicting and representing, together with quantifying, coordinating, and organising data" (English, 2004, p. 207). Having the above qualities does not guarantee one's success in the workplace. The ability to plan, monitor, and communicate results, together with being able to work collaboratively with other work partners are factors that lead to one's success in the workplace (Lesh and Doerr, 2003). This has an important implication for education systems; they must not only provide a strong education foundation for the students, they need to provide students with mathematical experiences that nurture and develop 21st century competency skills in the students. One suggested approach to do so is through providing students with mathematical modelling experiences in the mathematics curriculum.

14.2. Mathematical Modelling Process and Mathematical Modelling in Singapore

In his recent National Day Rally Speech, the Prime Minister of Singapore, Lee Hsien Loong, posed two problems to Singaporeans: One, how to preserve the green spaces along the Keretapi Tanah Melayu (KTM) railway line without affecting the development potential of the land? Two, to design a centre for bicycle activities where Singaporeans may go to rent bicycles and also have a meal on their bikes (Lee, 2011). All these are real-world problems and mathematical modelling can be used to conceptualise these problem situations.

So what is mathematical modelling? The term, "mathematical modelling" does not seem to have a universally agreed upon definition that mathematicians and mathematics educators agreed upon. Neither does there exist a "homogeneous understanding of modelling and its epistemological backgrounds" (Kaiser & Sriraman, 2006, p. 302).

In this chapter, we use the following definition of mathematical modelling:

> Mathematical modelling is a process of conceptualising and representing real-world problems in mathematical terms in an attempt to find solutions to the problems (Ang, 2001).

A mathematical model is used during this attempt to find solutions to problems. This mathematical model can be considered a simplification or abstraction of the real-world problem created in a mathematical form. That real-world problem is translated into a mathematical form to give meaning

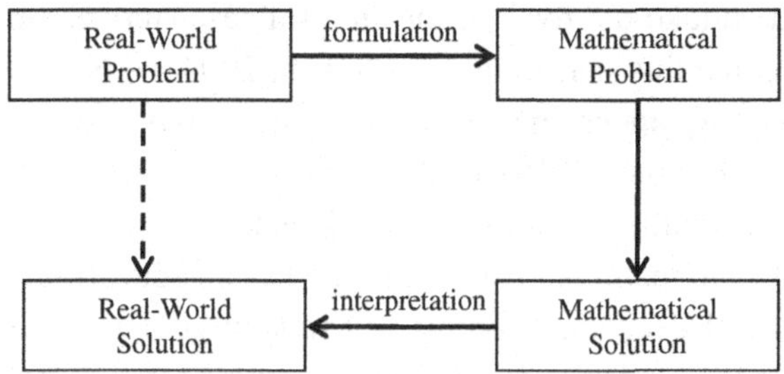

Figure 14.1. The mathematical modelling process (Ang, 2011).

to the conceptualised representations. The problem solver can then use this mathematical form to find solutions to the problem. The problem solver then uses whatever known techniques to obtain a mathematical solution. This mathematical solution is them interpreted and translated into real-world solutions.

Figure 14.1 shows the above described mathematical modelling process.

Proponents of mathematical modelling have called for the need to relook and rethink the nature of mathematical experiences provided in schools, and to include mathematical modelling in the school curriculum, in particular, in the primary school curriculum (see Chan, 2008; English, 2006; 2007). Their call stemmed from the concern that schools may not be providing adequate attention to help students develop 21st century competencies — abilities that are needed for success beyond school. They argued that traditional mathematics problems often consist of tasks that have restricted problem contexts where students just "generally engage in a one- or two-step process of mapping problem information onto arithmetic quantities and operations" (English, 2007). Specifically, the "inadequacy" in a traditional mathematics problem was that students were required only to be engaged in "situations where the 'givens', the 'goals', and the 'legal solution steps' were already specified in the problem" (English, 2004, p. 208). Hence "the interpretation processes for the child have been minimised or eliminated" (English, 2004, p. 208) in the sense that the problem has already been "mathematised" for the student who just needs to work from the "givens" to the "goals". The traditional problem does not give students opportunities to generalise and re-apply their learning. In contrast, mathematical modelling activities, which take the form of meaningful case studies, provide a solution to "overcome this inadequacy" (English, 2004, p. 208). Mathematical modelling not only opens an avenue for students to develop their ability to work with data (English, 2010), it also provides opportunities for students to "elicit their own mathematics as they work on the problem" (English, 2007, p. 30). In addition, mathematical modelling has an educational significance as it "promote[s] deep questioning and reflection" (Ikeda and Stephens, 2003), "enhances students' mathematical thinking in a creative manner" (Cheah, 2008) and also involves applying "mathematics which is useful in society" (Yanagimoto, 2005. p. 2).

Few research studies on mathematical modelling in Singapore primary and secondary classrooms have been conducted. In Chan's (2008) study, he observed that most primary school teachers associated mathematical modelling to the model approach. The model-drawing approach is a problem-solving method commonly taught in the Singapore classrooms to help students visualise and solve problems. In this approach, bar models are used to represent mathematical quantities (known and unknown) and their relationships given in a problem. This erroneous association of mathematical modelling to the model approach, as Chan noted, is no surprise as mathematical modelling has very much been seen, until recently, as suitable for secondary grade levels and above.

In another study, Ng (2011a) found that Secondary 1 and 2 (aged 13–14) students "did not apply all the expected mathematical knowledge and skills afforded by the [mathematical modelling] tasks" (p. 115) in spite of them already having these knowledge and skills. In addition, these students were observed to display "limited activation of real-world knowledge in mathematical application and decision-making" (p. 115). Teachers could help their students by providing some scaffolding. However, skill is required on the part of the teacher to provide just the right amount of scaffolding to their students. Indeed, it is a challenge for teachers to find a "balance" between scaffolding students and relinquishing control by giving students opportunities to discover and engage in their own mathematical solutions to the mathematical modelling tasks (Chan, 2008; Ng, 2011a).

Being mindful of the importance of providing students with mathematical experiences that nurture and develop 21st century competency skills (see Chapter 4.6 for the Singapore Mathematics Curriculum and its connections with mathematical modelling), the Singapore Ministry of Education also recognises that knowledge and 21st century skills must be underpinned by values — which shape a person's beliefs, attitudes and actions. Such values in turn define a person's character. The values at the core of the 21st century competencies, as stipulated by the Singapore Ministry of Education are: respect, responsibility, integrity, care, resilience and harmony (Ministry of Education, 2010). This fits perfectly in with the spirit of mathematical modelling as "problem solving is seldom limited as a mission for an individual to work on because of issues, ambiguities and the lack of complete information (Chan, 2008, p. 52). Hence through working collaboratively with their fellow teammates, students are given the opportunity to develop socially as in a team, they have a shared responsibility to problem resolution.

14.2.1. *The 4-man tent problem*

Five groups of Primary 6 students (aged 11–12 years old) and five groups of Secondary 1 students (aged 12–13 years old) participated in the Mathematical Modelling Outreach event held in Singapore in June 2010. The Mathematical Modelling Outreach was organised by the National Institute of Education (NIE), the sole teacher-training institution in Singapore. Each group consisted of three or four students. There was one primary and one secondary group students which consisted of students from an Australian school. Two NIE pre-service teachers were in-charge of facilitating the primary cohort. Similarly, the secondary cohort also had two NIE pre-service teachers as their facilitators.

The modelling task was a "Design a 4-man tent" problem (see Figure 14.2).

The framework used for facilitating the modelling process and scaffolding student learning consists of six phases. The six phases were: Discuss, Plan, Experimentation, Verification, Product and Reflection.

Each group of students was given a booklet to work on the task. Each booklet consisted of the problem itself and six sections to guide students as they worked on the problem. The six sections reflected the abovementioned mathematical modelling process framework, that is; Discuss, Plan, Experimentation, Verification, Product and Reflection.

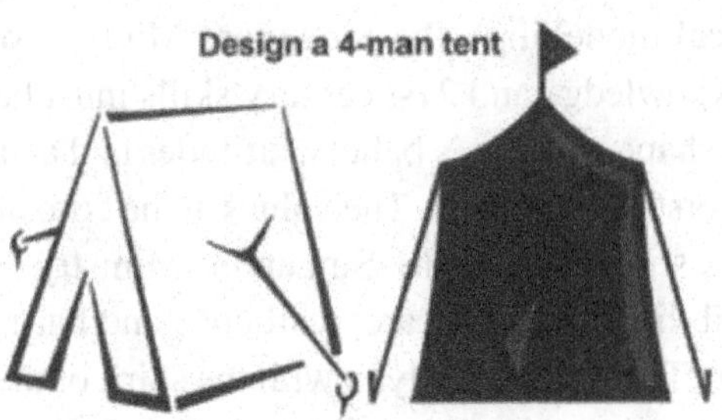

As temperatures around the world see an increase due to global warming, more people are also going to the beach to cool off. Adventurous Spirit, a company manufacturing outdoor equipment for adventures and sports, realises that they could produce and sell tents to beach-goers. They needed to come up with appropriate designs for tents that are saleable and would be popular with the beach-goers. They have invited your team to design a 4-man tent and make your proposal to them.

Figure 14.2. The "Design a 4-man tent" problem.

14.2.2. *Results*

This section presents four of the six phases of the modelling processes displayed by both the primary and secondary school groups — discussion, planning, experimenting and verifying.

14.2.2.1. Primary group: A documentation of their initial development of problem representation

Each primary school group got down to discuss the given problem with their teammates right after the two NIE pre-service teacher facilitators introduced the 4-Man tent problem to them. The early period of solving the problem saw the students trying to list down the features they would like to include in their 4-Man tent design. Table 14.1 shows the factors which each group of primary school students considered for building the 4-man tent. P01 stands for Primary group 1, P02 stands for Primary group 2, P03 stands for Primary group 3, P04 stands for Primary group 4 and P05 stands for Primary group 5.

As Group P01 had the most comprehensive consideration of factors for their tent design, henceforth, only their mathematical modelling processes will be presented and discussed.

Table 14.1. *Factors which each group of primary school students considered for building the 4-man tent.*

	P01	P02	P03	P04	P05
Shape	√	√		√	
Dimension	√	√	√	√	√
Weight	√				
Type of material	√	√			
Weather consideration	√	√		√	
Cost	√				
Comfort	√	√			
Colour	√			√	
Portable/easy to set up	√		√		

As shown in Table 14.1, Group P01 thought and came up with the most number of features they would take into consideration in the design of their tent. Their first consideration was the dimensions of the tent.

S1: Our team's goal is to pitch a tent that can fit four people.
S3: The length of a tent, the height of a tent, the width of a tent ... must fit at least four people.
S2: What information do you need? The height of the person.

Group P01 also discussed about the shape of the tent and the type of material needed for the tent. They decided on a pyramid-shaped tent. They also agreed to use a cooling material for the tent as they were selling the tents to beach-goers. They also decided that the tent to be white in colour as white absorbs the least amount of heat compared to other colours. Cotton was decided for the base of the tent so that it is comfortable to sit on or lie on. Student S3 pointed out to his teammates that the weight of the tent had to be light as he was considering the portability of the tent. He also wondered whether they should consider the ease of setting up the tent.

14.2.2.2. Secondary group: A documentation of their initial development of problem representation

Each of the four secondary school group got down to discuss the given problem with their teammates right after the two NIE pre-service teacher

Table 14.2. *Factors which each group of secondary school students considered for building the 4-men tent.*

	S01	S02	S03	S04	S05
Shape	√	√	√	√	√
Dimension	√	√	√	√	√
Type of material	√	√		√	√
Weather consideration	√	√			√
Cost of material	√	√	√	√	√
Safe	√				√
Comfort	√				
Colour				√	
Portable		√			
Foldable (storage)	√				
Stable					√
Durable		√			

facilitators introduced the 4-Man tent problem to them. Similar to their primary counterparts, the early period of solving the problem saw the students trying to list down the features they would like to include in their 4-Man tent design.

Table 14.2 shows the factors which each group of secondary school students considered for building the 4-man tent. S01 stands for Secondary group 1, S02 stands for Secondary group 2, S03 stands for Secondary group 3, S04 stands for Secondary group 4, and S05 stands for Secondary group 5.

As shown in Table 14.2, all the secondary school groups took the shape, the dimensions of the tent and the cost of materials for the tent into their consideration of the tent design. As Group S01's written recording of their mathematical modelling processes was the most complete among the five secondary groups, discussion henceforth would focus only on this group of secondary school students.

Group S01 decided that their team's goal was to design a multi-purpose tent that is suitable for sleeping at the beach, for camping and it had to the aesthetically pleasing. They also required that their tent be easily foldable for easy storage. They explored the types of tent already available in the market and wondered whether they would be able to find information on the tent designs that customers prefer, and those design that customers would avoid.

14.2.2.3. Primary group: A documentation of their intermediate development of problem representation

As Group P01 had the most comprehensive consideration of features for their tent design, only their mathematical modelling processes: planning and experimenting will be presented.

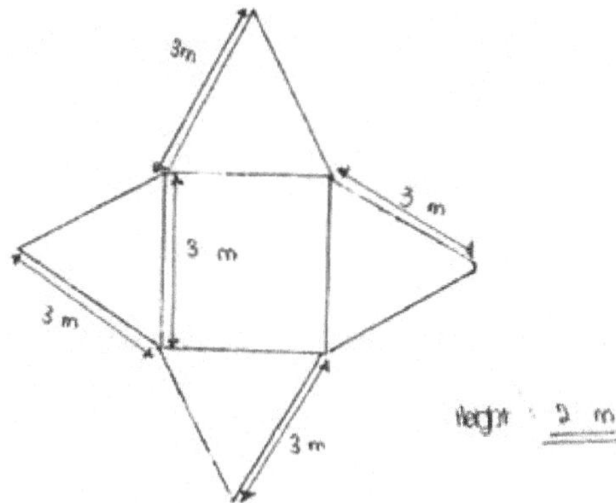

Figure 14.3. Group P01's design of the tent at the planning stage.

During the planning stage, the students were constantly checking on each other's ideas:

S2: First, we need to cut the canvas from a single piece.
S1: The canvas must be light-coloured and not too heavy.
S2: Yes, we also need to measure the right measurement (see Figure 14.3 for their mathematical model).
S3: Then we need some sticks, ropes or nails to hold the canvas firmly.
S1: Don't use so much, if not the tent would be heavy!
S2: On top of the tent, we can cut a part and replace it with a transparent strong plastic. So that people inside can see through, and it is waterproof so water will not sip through.
S3: We can put an opaque flip over the transparent part, so that we can close it during the day so that the sunlight would not hurt the eyes. We can open it at night and see the sky for our entertainment.
S1: Okay!

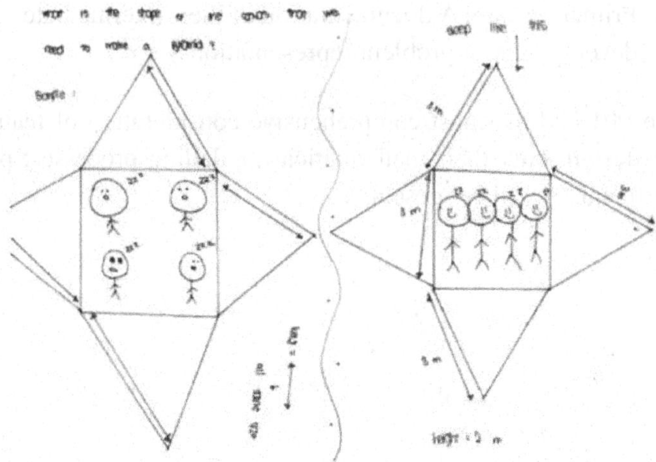

Figure 14.4. Group P01's design of the tent at the planning stage — fitting in four men into the tent.

During the experimentation stage, the students decided that the height of their tent to be 2 metres, so that a person with a height of 1.8 metres could stand up inside the tent comfortably.

They also thought about how to fit in 4 sleeping men into the tent. Figure 14.4 shows the two different ways that the students considered to position and fit 4 men into the tent. The students did not consider the scenario where if all the 4 people were of height 1.8 metres, then it would not be possible to fit them in the sleeping position shown on the left of Figure 14.4.

These students also noted in their booklet that they were not sure how to calculate the volume of their tent. The shape of their tent was a pyramid and they have not learnt how to compute the volume of a pyramid in their mathematics classes. They erroneously computed the volume of the tent as follows:

$$\frac{1}{2} \times 2 \times 3 = 3$$

$$3 \times 3 = 9\,\text{m}^2$$

Volume of the tent $= 9\,\text{m}^2$

14.2.2.4. Secondary group: A documentation of their intermediate development of problem representation

During the planning stage, the students in S01 decided to add extra features such as zips, a mosquito net and storage space inside the tent which they had. Figure 14.5 shows their initial conceptual representation of the problem. They decided that the height of their tent be 1.5 metres.

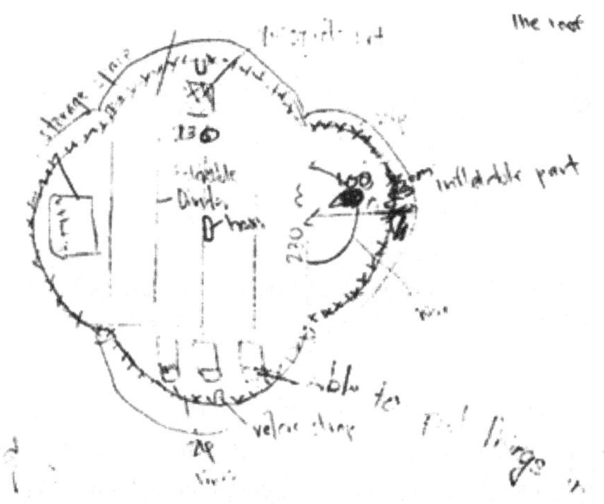

Figure 14.5. Group S01's initial design of the tent during the planning stage.

During the experimentation stage, they decided to change the shape of their tent to a pyramid (see Figure 14.6) as they realised that the initial shape of their tent was not viable — that the tent will not be enclosed. Unlike their primary counterparts, these students had no difficulty computing the volume of their tent.

14.2.2.5. Primary group and Secondary group: A documentation of their verification of their problem representation

The following shows what the primary Group P01 wrote in their booklet for their verification of their problem representation:

Our design is creative and it is cheap. It can fit four people and there is extra space. The extra is about one-quarter of the whole tent. No, it is

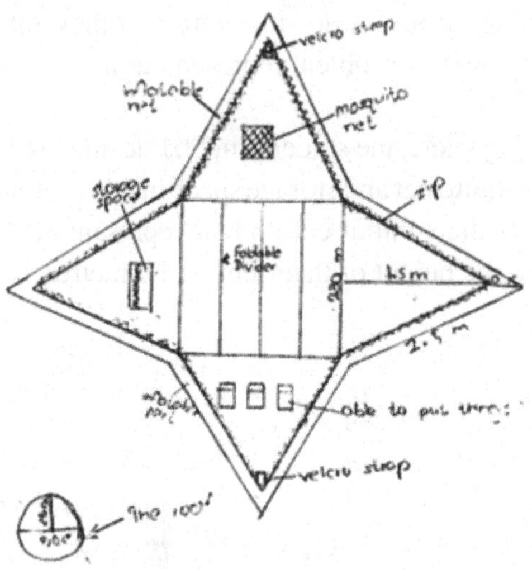

Figure 14.6. Group S01's amended design of the tent.

not easy to draw and cut out the net of the tent as we had to measure every thing we did; to ensure accuracy. Our tent is tall enough. It is 2 m tall.

The following shows what the secondary Group S01 wrote in their booklet for their verification of their problem representation:

Our sleepy zippy tent can be for many purposes. It can be used for sleeping, camping and storage. It is very suitable especially for beach-goers because it is very light-weight and foldable. Beach-goers can easily build a tent and fold it when they keep. The unique thing about this tent is that its walls can zip up. People can just open the tent with the zip. There are also inflatable parts attached together with the walls to make the tent stable. They can be pumped by air. The total cost of the tent is very reasonable so the sleepy zippy tent is the ideal tent for people especially beach-goers.

14.3. Discussion

As presented in Section 14.2.2, there are obvious differences between the modelling processes of the primary school and secondary school groups.

While both P01 and S01 groups took into consideration the basic requirements of designing a tent (dimensions of tent, weather considerations etc.), the secondary school students took a further step to ask what tent designs were already in the market. They also took into consideration the tent designs consumers would like prefer to have and those design that consumers would avoid buying.

English (2006) gave six reasons to argue for the inclusion of mathematical modelling in the school curriculum, in particular, the opportunity for optical mathematical development where students are engaged in important mathematical processes. Both the primary and secondary groups of students were observed to be engaged in some mathematical analysis and reasoning, as they communicate their logical arguments to the other team members. However, there is room for more mathematical rigour in the mathematical thinking, reasoning, arguments and documentation (See Section 14.2.2.5) of their verification of their model for both primary and secondary groups. For example, there were a number of features of the tent which the students included in their tent design and not mathematically accounted for. In the primary P01 group, the dimensions of the transparent viewer and the opaque cover for the viewer, and the weight of the tent were not presented in their mathematical arguments. In the secondary S01 group, the zips, foldable parts and inflatable parts of the tent were casually mentioned but not accounted for mathematically. As mentioned in Section 1.2, Ng (2011a) also reported that students were found to be lacking in mathematical rigour in their mathematical application and decision-making.

The reason for this lacking in mathematical rigour could be due to the following: after finding a mathematical solution to the mathematical model, the students did not proceed to re-interpreting and translating their mathematical model into real-world solutions. Indeed, this step was observed to be missing in both the students' verifications and reflections — for both primary and secondary groups.The cost of manufacturing the tents were considered by both the primary and secondary groups of students. However, they did not focus on this essential consideration. In real life, designers need to constantly remind themselves not to "lose sight of the bigger picture" (Petroski, 2011, p. 291) and forget to keep the price of the product affordable. Another observation was that; had Group P01 translated their mathematical solution into a real-world solution, they would have discovered that their tent, which is of height 2 metres, would not only add to the cost but also the ease of portability and setting up of the tent.

On a positive note, the observations in terms of how the students responded and worked on the "Design a 4-man tent" problem are encouraging. This is perhaps the students' and NIE pre-service teacher facilitators' first formal encounter with mathematical modelling. No doubt, there is still room for improvement for these students' depth of mathematical reasoning when verifying their mathematical model. After all, the most challenging part of mathematical modelling for these students was not to simply present a wish-list of what features they would like to have in their tent design, but how to make practical decisions and convincing arguments for their wish-list through mathematical thinking and reasoning. All these provide implications for teaching.

14.4. Implications and Conclusion

Chan (2008) cautioned that promoting mathematical modelling in the primary mathematics classrooms "can be quite a challenge as teachers need to make sense what those [mathematical modelling] processes mean, how they 'look like' and how and by what means they can be developed" in the students (p. 61). Indeed, teachers need to be given both time and professional training to get in tune with this non-traditional teaching (mathematical modelling) mind set. Asking the appropriate probing questions is "essential in the teaching of mathematical modelling" (Ikeda and Stephens, 2003, p. 211). Ng (2011b) aptly suggested the development of two key competencies when teachers go for mathematical modelling training — developing the appropriate questioning skills, when and how to ask 'diagnostic' questions to help students on their approaches and thinking and to assess students' progress, and how to foster a modelling climate in the classroom.

The author would also like to suggest providing teachers with professional training on how to identify and choose both meaningful and interesting real-life mathematical modelling tasks for their classroom teaching. Besides the "Design a 4-man tent" task, another example of such a task is how to fit a round pizza into a square pizza box without having the top of the box touching the pizza's cheese (see Petroski, 2011). The Ministry of Education could also help by giving clearer guidelines for the inclusion of mathematical modelling in the mathematics curriculum, especially in the primary curriculum, so as to help teachers better prepare their students for the 21st century workplace.

References

Ang, K. C. (2001). Teaching mathematical modelling in Singapore schools. *The Mathematics Educator*, 6(1), 62–75.

Chan, C. M. E. (2008). Using model-eliciting activities for primary mathematics classrooms. *The Mathematics Educator*, 11(1/2), 47–66.

Cheah, U. H. (2008). Introducing mathematical modelling to secondary school teachers: A case study. *The Mathematics Educator*, 11(1/2), 21–32.

English, L. D. (2004). Mathematical modelling in the primary school. In I. Putt, R. Faragher, & M. McLean (Eds.), Mathematics education for the third millennium: Towards 2010 (pp. 207–214). James Cook University: Mathematics Education Research Group of Australasia.

English, L. D. (2006). Mathematical modelling in the primary school: Children's construction of a consumer guide. *Educational Studies in Mathematics*, 63, 303–323.

English, L. D. (2007). Interdisciplinary modelling in the primary mathematics curriculum. In Watson, Janw and Beswick, Kim (Eds.), Proceedings of the 30th Mathematics Education Research Group of Australasia Annual Conference (pp. 275–284). Hobart.

English, L. D. (2010). Young children's early modeling with data. *Mathematics Education Research Journal*, 22(2), 24–47.

Lesh, R. A. & Doerr, H. (2003). Foundations of a models and models perspective on mathematics teaching and learning. In R. A. Lesh & H. Doerr (Eds.), Beyond constructivism: A models and modelling perspective on mathematics teaching, learning and problem solving (pp. 3–34). Mahwah, NJ: Erlbaum.

Jones, A. (2010, October 9–10). Good jobs have never heard of. *The Weekend Australian*, p. 8.

Ministry of Education (2010). Nurturing our young for the future: Competencies for the 21st century. Singapore: Ministry of Education.

Ng, K. E. D. (2011a). Mathematical Knowledge Application and Student Difficulties in a Design-Based Interdisciplinary Project. In G. Kaiser, W. Blum, R. Borromeo Ferri & G. Stillman (Eds.), *International perspectives on the teaching and learning of mathematical modelling: Trends in the teaching and learning of mathematical modelling* (Vol. 1, pp. 107–116). New York: Springer.

Ng, K. E. D. (2011b, 14–19 Jul 2011). Facilitation and scaffolding: Symposium on Teacher Professional Development on Mathematical Modelling–Initial perspectives from Singapore Paper presented at the Connecting to practice — Teaching practice and the practice of applied mathematicians: The 15th International Conference on the Teaching of Mathematical Modelling and Applications, Australian Catholic University (St. Patrick), Melbourne, Australia.

Lee, H. L. (14 August, 2011). National Day Rally 2011. Speech given at University Cultural Centre, National University of Singapore, Singapore.

Ikeda, T. & Stephens, M. (2003). Comparing an analytical approach and a constructive approach to modelling. In S. J. Lamon, W. A. Parker & K. Houston (Eds.), *Mathematical modelling: A way of life — ICTMA* 11, (pp. 202–211). West Sessex, UK: Horwood Publishing Ltd.

Kaiser, G. & Sriraman, B. (2006). A global survey of international perspectives on modelling in mathematics education. *ZDM*, 38(3), 302–310.

Petroski, H. (2011). A round pie in a square box. *American Scientist*, 99(4), 288–292.

Yanagimoto, T. (2005). Teaching modelling as an alternative approach to school mathematics.*Teaching Mathematics and Its Applications*, 24(1), 1–13.

CHAPTER 15

LEARNING THROUGH "DREAM HOME"

CHAN Chun Ming Eric

Are children capable of designing? This chapter describes how two groups of Primary 5 pupils worked on a mathematical modelling task in designing a 5-room flat. The modelling task was structured in a rich context that required the pupils to apply their curricular knowledge involving geometry and measurement towards coming up with an ideal floor plan. The modelling process describes the various conceptualisations the pupils conceived as models. As designers, they were capable of mathematising and justifying why they designed the floor plan to be as such. Finally, implications are discussed with respect to using mathematical modelling as problem-solving to support mathematics education reformed efforts.

Keywords: Primary School, Measurement, Mathematical Reasoning, Model Development, Modelling Process.

15.1. Introduction

In this mathematical modelling activity, Primary 5 pupils are seen as designers. From a models-and-modelling perspective (Lesh & Doerr, 2003), pupils develop useful conceptions expressed as thought-revealing representations that are continually tested and revised towards meeting some specified purposes. The modelling process that the pupils go through results in the development of models through mathematising the situation. In other words, the pupils are engaged in interpreting problem information, selecting relevant quantities, identifying operations that leads to new quantities, and creates meaningful representations (English, 2006; Lesh & Doerr, 2003). The mathematics that they develop gives meaning to their conceptualisations. In this sense, the models appear as conceptualised designs and the mathematical relationships expressed as descriptions, drawings, and calculations. This chapter presents the mathematical modelling endeavours of two groups of Primary 5 pupils in the design of a 5-room flat.

15.2. The Modelling Task

The task that the pupils were engaged in was entitled "My Dream Home". The main context is shown in Figure 15.1a. Figure 15.1b shows the requirements the pupils had to work within.

The modelling task presents pupils with a meaningful and authentic context to work on by designing a floor-plan of a 5-room flat and furnishing the master bedroom given a certain budget. The design of the task was based on the criteria of modelling design principles for reality, model construction, self-evaluation, documentation, prototyping, and model generalisation (Lesh *et al.*, 2003).

> Noble Construction Designers are hiring "architects" to (1) design and draw the floor plan for a 5-room flat and (2) propose how the bedroom can be furnished within the given budget. Your team will be working on the design and furnishing ideas. The drawing has to be properly labeled, stating the various parts of the flat and its real-life dimensions. Furnishings for the bedroom include furniture items. Present and explain to Noble Construction Designers to justify your design.

Figure 15.1a. The modelling task.

> Requirements given by Noble Construction Designers:
>
> A 5-room flat should comprise
> four bedrooms
> a living room
> two washrooms/toilets
> a kitchen
> a storeroom
> a small balcony
> total floor area 132 m^2
>
> 2. The floor plan must be able to fit into the size of the plastic hardboard given.
> 3. The floor plan drawing should represent the real-life dimensions of the 5-room flat.
> 4. The total cost of furnishing for the bedroom should not exceed $3000.
> 5. Use only the Ikea catalogue provided to choose your furniture items.

Figure 15.1b. Requirements of the modelling task.

Where the mathematics aspects are concerned, the pupils would have to manage specific mathematical concepts involving geometry, measurement, estimation and costing. For example, they need to make considerations like how the areas of the rooms could be drawn to scale, the shapes of the rooms, the realistic size a certain type of room space that is required, and even the aesthetics aspects as well. In so doing, pupils make visible their mathematical thinking and mathematising through their descriptions, explanations and justifications that are integral to the models they develop. As pupils have to work together in small groups of three or four, it also shows that learning involves a social dimension as stemming from a socialcultural perspective. The cycles of testing and revising that arise when pupils confront the task and their friends through the application of prior knowledge, correcting knowledge, new knowledge and connecting knowledge are what is deemed as constituting understanding during mathematical modelling (Moschkovich, 2004).

15.3. Exemplification of Pupils' Modelling Conceptualisations

The mathematical modelling process in this activity comprises five subprocesses:

(i) Discuss — Pupils to discuss goals, information they need, and how to use the information.
(ii) Plan — Pupils to make a clear record of things they need to do; sketch their drafts.
(iii) Experimentation — Pupils show their calculations.
(iv) Verification — Pupils make final recommendations and plan their justifications.
(v) Product — Pupils present their product.

Two groups of pupils' conceptualisations based on the above process will be presented to show how they accomplish the modelling task. Group A comprised three pupils from the same school and Group B comprised three pupils from different schools.

<u>*Discuss*</u>

Both groups began with discussing what they hoped to accomplish. Their discussion of goals and information given or needed was meant to scaffold their planning later on. For Group A, they stated that their goals were to plan and design a 5-room flat, and to furnish the master bedroom. Group B was more explicit in that they specified the condition — "*Using \$3000 to furnish the master bedroom*" suggesting their awareness of task conditions.

In this stage, the pupils interrogated the task details to surface important information. For example, Group A wanted to work on "*the size of the flat*". Group B went deeper by stating "*the necessary items for the master bedroom*", "*how big is 132 m²*", "*shape of the house*" and "*size of rooms*". The initial design became more concrete as they began to plan by making a record of what they needed to do.

<u>*Plan*</u>

Both groups made a list of things they needed to do. The plans saw them begin to initiate ideas that involve mathematisation. These ideas are presented in Figure 15.2 below.

A draft sketch of Group B's 5-room flat is shown in Figure 15.3. Before arriving at the sketch, the pupils made use of measuring tapes to measure actual room and door sizes to get a spatial sense of the dimensions of things in the environment.

The sketch shows the pupils' conceptualisation of the rooms as rectangles; typical of rooms in Singapore flats. Different types of rooms as stated in the task details like living room, bedroom, store room and kitchen are conceived and positioned in the design. The pupils also used their personal knowledge to include a laundry area and study corners. Room size

Group	What the groups wrote	Mathematisation Aspect
A, B	*design floor plan; sketch of the house plan; draw a sketch of the furnished master bedroom*	Designing layouts
A, B	*find the area of the floor plan; estimate the area of each room*	Qualifying mathematical components needed
B	*draw the dimensions of the house*	Dimensionalising
A	*find a place to put the furniture*	Positioning
A	*make it such that 2 cm is 1 m and labelled*	Scaling
A, B	*spend all the money; choose furniture from the catalogue*	Costing

Figure 15.2. Planning and mathematising.

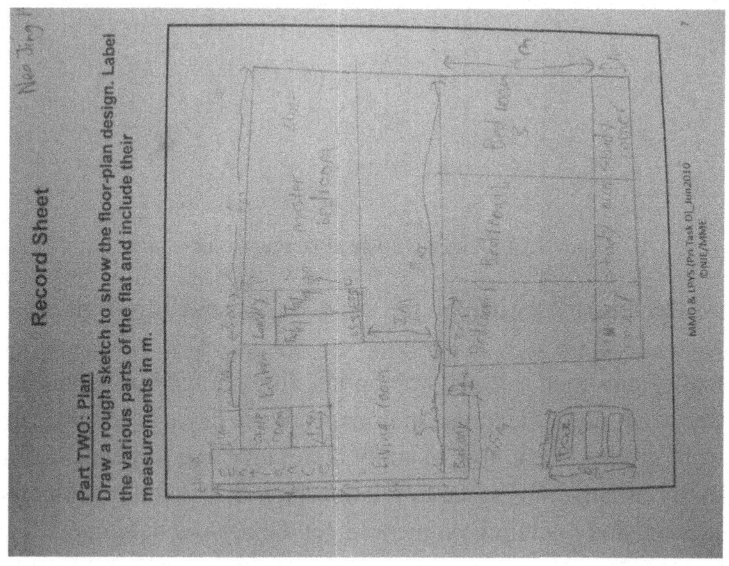

Figure 15.3. Group B's sketch of a 5-room flat.

and dimensions appeared reasonable but they did not suggest meeting the condition of the entire floor area being 132 m^2 at that juncture. Drawing to scale is not apparent as well. Nonetheless, this initial model paved the way for improving the model further.

Experimentation

Getting a floor area of 132 m^2 was anticipated to be difficult as groups were suspected to use a trial method to sum up rectangular areas for that purpose. Group B indeed used the trial method. Group A, however, worked creatively to get the required area. They drew a rectangle measuring 13 cm by 12 cm which constituted an area of 156 cm^2 (Note that the pupils used "cm" to represent "m" in their drawings). To get 132 cm^2, they needed to just subtract 24 cm^2 from 156 cm^2. With 132 cm^2 confirmed (treated as analogous to 132 m^2), they were assured that the floor area would be correct no matter how they partitioned the rooms within the parameters. Following this reasoning, they went on to try reasonable dimensions of the rooms and their positioning (see Figure 15.4).

Group A then prepared their drawing of the floor plan through a scale where 2 cm represented 1 m. Subsequently, further revisions were made to

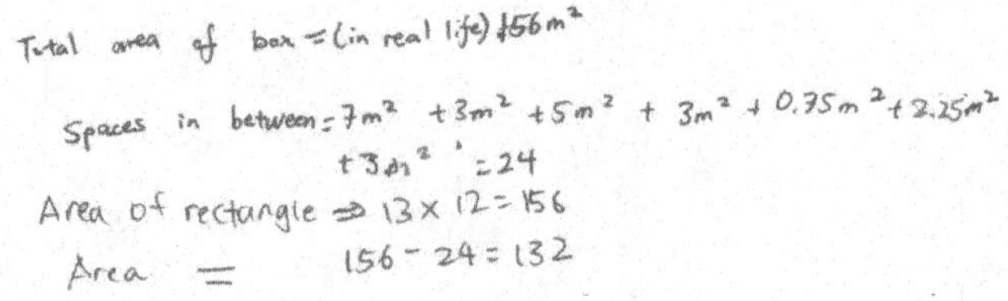

Total area of box = (in real life) 156m²

Spaces in between = 7m² + 3m² + 5m² + 3m² + 0.75m² + 2.25m² + 3m² = 24

Area of rectangle ⇒ 13 × 12 = 156

Area = 156 − 24 = 132

Figure 15.4. A creative way of getting the floor area.

improve the drawings whereby the scale was changed to 4 cm representing 1 m.

Both groups then selected appropriate furnishings for the master bedroom and went on to maximise their expenditure close to $3000. Through it all, both groups learnt how to apply their curricular mathematics meaningfully by establishing variable relationships that involve geometry and measurement and operating on them to get the desired outcome.

Verification and Product

The testing and revising of their models must come to a point where pupils were satisfied that their model would meet the goals that they had set out to achieve. Their models must reflect the realistic situation of meeting the requirements of the company (Noble Construction Designers). The pupils shared their designs with fellow pupils and teachers as they communicated their ideas. In this respect, this part of the modelling process became a reflection session where constructive feedback on one another's model was attained.

Group A's final model (see Figure 15.5) is used to exemplify the groups' verification of their models, and their written records are used to give meaning to what they see as strengths in their model. As seen, the final floor plan was "professionally" drawn, very much like the real thing that architects would develop. Colours were used to differentiate the type of rooms and scaled dimensions were appropriately labelled. The layout met the design requirements as given in the task sheet.

One of Group A's design strengths, as noted by the pupils, lies in the huge toilets — "*The design is very appealing. Many people can use the toilet at one time. In the morning, nobody needs to compete for the bathroom*"; a realistic issue to be tackled, based on their personal experiences. The other strength they cited, was a corridor they had created. The purpose was that

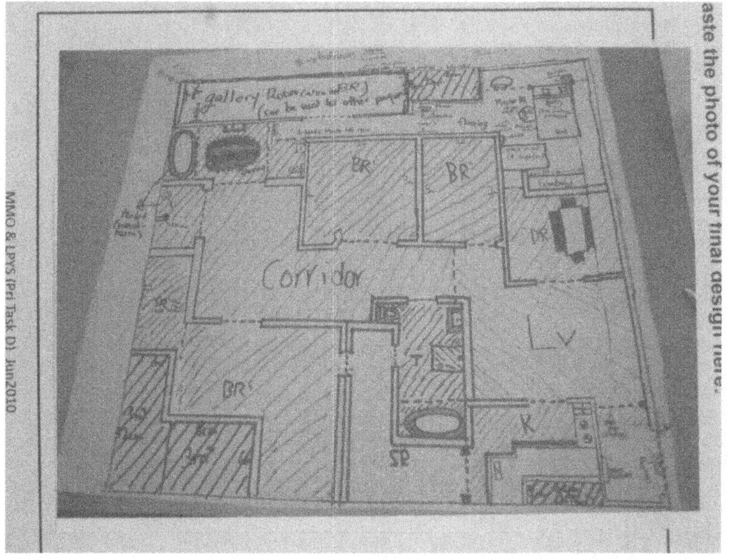

Figure 15.5. Group A's final floor plan.

"*when there is a rainy day, you can play catching . . . a lot of other games and sports indoor in the corridor. This will satisfy youngsters*".

Indeed modelling from a child's point of view takes into consideration how they themselves fit into the picture. The pupils also thought about convenience and connectivity when they wrote, "*The master bedroom's walkway is also connected to the bathroom and bathroom 3 is also in front of the second bathroom, thus the person using bathroom 3 only needs to walk straight to (get to) the toilet.*"

Group B's final model is shown in Figure 15.6. Their design layout is similar to that of their draft in Figure 15.3 except that now it is drawn to scale and the master bedroom is furnished. The rooms that make up the floor plan adhere to the task conditions given.

Group B's perception of the strengths of their design was vastly different from Group A's. They had an entrance corridor which they claimed can "*prevent guests outside from seeing what is in the living room*" which in a sense, suggests there is minimal invasion of privacy from outsiders. They factored additional rooms that were not stated in the task sheet. These include the laundry room and the study corner. They wrote, "*The laundry*

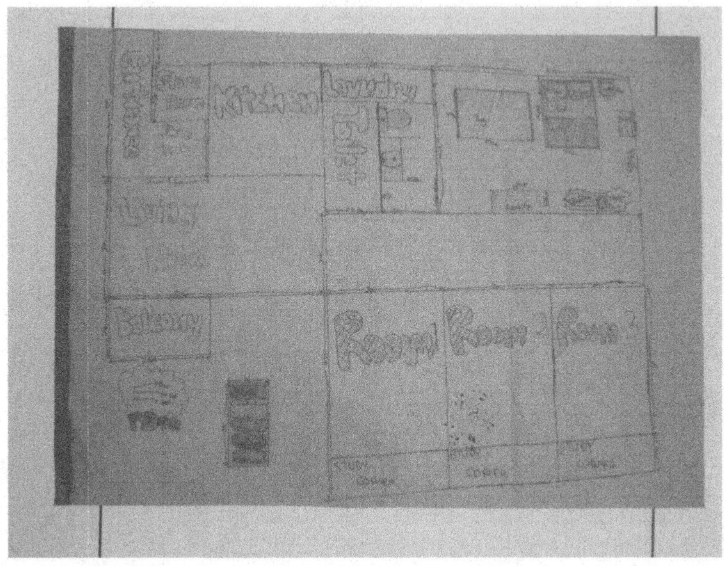

Figure 15.6. Group B's final floor plan.

is the perfect place to dry clothes." and "*The laundry is close to the kitchen because the heat in the kitchen can raise the temperature of the laundry so the clothes can dry faster.*" suggesting good positional and practical sense. Where the study corner is concerned, they claimed, "*A study corner is just the right place to study.*" and "*The study corner must be separated from the room as it will be more private.*" This group apparently had considered deeper into the basic needs of people living in a flat than what had been stipulated in the task sheet.

15.4. Implications

The mathematical modelling process takes the pupils through various modelling sub-processes such as discussing, planning, experimenting and verifying. Within each sub-process, the work of mathematising is involved. Pupils learn how to frame the problem and make meanings of their conceptualisations through establishing relationships between variables and mathematical reasoning. In this sense, pupils develop emerging models and work towards refining their models. Such an endeavour is valued because it provides an authentic situation for pupils to construct their own mathematics to solve the problem (English, 2009). It also provides them the opportunity to apply the curricular mathematics they have learnt in the classroom that they seldom could make use of in new situations (Chan, 2009). As seen, their application of mathematical knowledge is not compartmentalised. The mathematical reasoning tends to be supported with their personal knowledge or experiences and the decisions to make under certain circumstances. Such integration is seen as essential to developing mathematical power (Katsberg, D'Ambrosio, McDermott & Saada, 2005).

The design of the modelling task based on modelling principles essentially helps to distinguish it from other types of problem-solving tasks that do not elicit a high level of mathematising cum social-cognitive engagement. To make the activity more intentional in its approach, a goal-directed strategy is used much like in a problem-based learning setting. Getting pupils to think about and work towards their goals is seen as a form of scientific reasoning that helps the pupils to learn how to frame questions and implement strategies to answer those questions (Waters & Schneider, 2009).

The mathematical modelling endeavours of the pupils as shown, translates to a new dimension in the teaching of mathematics where pupils take more ownership of their learning. While it does take more time to implement and monitor, such activities should be integrated within existing classroom problem-solving sessions to avoid making it as a separate enrichment activity. Moreover, mathematical modelling should also be adopted as one that is supportive of having pupils to explore, explain, revise, justify and reflect on their ideas, and where getting stuck is not towards thinking deeper or looking for alternative ways (Lesh & Zawojewski, 2007).

15.5. Concluding Remarks

Mathematics education reforms are calling for meaningful and authentic ways to engage pupils in solving problems. Mathematical modelling is seen as a promising approach to help achieve that goal. This chapter has given a glimpse of what two groups of Primary 5 pupils were capable of when given a complex modelling task. Not only were they highly engaged, they manifested their mathematical thinking through mathematisation as they give meanings to situations.

References

Chan, C. M. E. (2009). Mathematical modelling as problem solving for children in the Singapore mathematics classroom. *Journal of Science and Mathematics Education in Southeast Asia*, 32(1), 36–61.

English, L. D. (2009). Promoting interdisciplinarity through mathematical modelling. *ZDM: The International Journal on Mathematics Education*, 41(1), 161–181.

Katsberg, S., D'Ambrosio, B., McDermott, G. & Saada, N. (2005). Context matters in assessing students' mathematical power. *For the Learning of Mathematics*, 25(2), 10–15.

Lesh, R., Cramer, K., Doerr, H. M., Post, T. & Zawojewski, J. (2003) Model development sequences. In R. Lesh & H. Doerr, (Eds.), *Beyond constructivism: Models and modeling perspectives on mathematics problem solving, learning and teaching* (pp. 35–58). Mahwah, NJ: Lawrence Erlbaum Associates.

Lesh, R. & Doerr, H. (2003). Foundations of a models and modeling perspective in mathematics teaching and learning. In R. Lesh & H. M. Doerr (Eds.), *Beyond constructivism: Models and modeling perspectives on mathematics problem solving, learning and teaching* (pp. 3–34). Mahwah, NJ: Lawrence Erlbaum Associates.

Lesh, R. & Zawojewski, J. (2007). Problem solving and modeling. In F. K. Lester (Ed.), *Second handbook of research on mathematics teaching and learning: A project of the National Council of Teachers of Mathematics* (pp. 763–803). Charlotte, NC: Image Age Publishing.

Waters, H. S. & Schneider, W. (2009). Common themes and future challenges. In H. S. Waters & W. Schneider (Eds.), *Metacognition, strategy use and instruction* (pp. 281–288). New York: Guilford Press.

CHAPTER 16

LEARNING THROUGH "THE UNSINKABLE TITANIC"

JAGUTHSING Dindyal

FOO Him Ho

This chapter describes how some Primary 5 pupils from Singapore interpreted, carried out and presented a paper-folding mathematical modelling task over a period of three days during a Mathematical Modelling Outreach. The students who worked in groups of 3 had to design and fold a paper boat that could carry different loads. The data shows that the students were deeply engaged in the modelling task and used several mathematical concepts to come up with a workable design. We discuss the implications of such tasks for the Singapore mathematics curriculum which is centred on problem solving.

Keywords: Primary School, Mathematisation, Mathematical Modelling Outreach.

16.1. Introduction

Problem solving has been at the core of the Singapore Mathematics Curriculum Framework (SMCF) for the past two decades. Singaporean teachers vary in their styles and approaches to problem-solving instructions in terms of the amount of class time and attention spent in the various classroom activities, and differing emphasis on each of the four problem stages of Pólya (see Hedberg, *et al.*, 2005). Also, on the international scene, there is now a growing concern that problem-solving research has not significantly informed school practice (see Lester & Kehle, 2003). Some authors have been quite critical about problem-solving approaches based on the works of Pólya and Schoenfeld. For example, English, Lesh and Fennewald (2008) wrote: "... whether attention focuses on Pólya-style heuristics or on Schoenfeld-style metacognitive processes or beliefs, short lists of descriptive processes or rules tend to be too general to have prescriptive power" (p. 3). The authors added that there is a flawed belief that students must first learn concepts and processes as abstractions before they can put them together and use them in "real-life" problem-solving situations. Thus problem solving tends to end up never getting taught at all in many classrooms. Clearly, the criticisms are not directed at problem solving itself. As an important process in mathematics education, it is still very highly valued. However, what it means to solve a problem and how to approach problem solving in the classroom are debatable. For example, Lesh and Zawojewski (2007), favouring the modelling perspective, claimed that we need to know how students interpret problem situations, how they mathematise them, how they quantify them and how they operate on quantities.

The latest version of the Singapore curriculum framework has a new addition namely, *Applications and Modelling*, which has been included within the *Processes* component of the SMCF. The 2006 document from the Curriculum Planning and Development Division (CPDD) strongly emphasises the importance of applications and modelling:

> *Application and Modelling play a vital role in the development of mathematical understanding and competencies. It is important that students apply mathematical problem-solving skills and reasoning skills to tackle a variety of problems, including real-world problems. (CPDD, 2006, p. 8)*

Applications and modelling are not located within specific content domains but rather it is expected that they cut across different areas in the curriculum. The document does not provide much support for the average teacher about how to implement applications and modelling in the classroom. However, some details about what is meant by mathematical modelling and what are some the learning outcomes are highlighted in the document.

In this chapter, we report how Primary 5 students worked on a "real-life" problem solving task in a Mathematical Modelling Outreach programme. More specifically, we look at how the students approached the task and what were some of the implications for integrating such tasks in the Singapore mathematics curriculum which is focused on problem solving.

16.2. Modelling

Modelling is involved when there is the need to create a model to capture the essence of some piece of reality and subsequently, to manipulate the model to arrive at some conclusions about the piece of reality. Swetz and Hartzler (1991) described mathematical modelling as a systematic process that draws on many skills and employs the higher cognitive activities of interpretation, analysis and synthesis. The authors suggested that the modelling process is composed of four main stages:

(1) Observing a phenomenon, delineating the problem situation inherent in the phenomenon, and discerning the important factors (variables/parameters) that affect the problem.
(2) Conjecturing the relationships among factors, and interpreting them mathematically to obtain a model of the phenomenon.
(3) Applying appropriate mathematical analysis to the model.
(4) Obtaining results and reinterpreting them in the context of the phenomenon under study and drawing conclusions. (p. 2)

This framework can be used to look at students' work on an authentic problem solving task at primary level.

16.3. The Mathematical Modelling Outreach

The Mathematical Modelling Outreach (MMO) was organised during the mid-year school break in Singapore with the twin objective of introducing mathematical modelling to primary students and also providing teachers with some ideas about how to implement modelling activities in the classroom. Four students and an accompanying teacher from 30 schools (28 local and 2 international schools) participated in MMO over a period of three days. Students were grouped according to their levels (primary or secondary) and had to work on specific modelling tasks selected for them. The "Unsinkable Titanic" modelling task (See Figure 16.1) was carried out by 12 students coming from 3 different primary schools. In Groups A, B and C, the 3 students in each group came from the same school whereas in Group D, the students came from three different schools. The task was carried out

Task Sheet

The Unsinkable Titanic

Craftworks, a book publisher, wants to produce a book for hobbyists and children to make different origami crafts. Apart from designs of aeroplanes and animals, they are specifically looking to design a boat that children can make and yet able to carry loads knowing that children like to put things inside a boat.

Craftworks have invited your team to design various types of boats and find out which design works best and make your recommendations to them.

Figure 16.1. The modelling task — the unsinkable titanic.

over a 3-day period during the MMO organised by the National Institute of Education (NIE), Singapore in June 2010. Two NIE undergraduate students were the facilitators throughout the whole duration under our supervision.

The materials that were provided and could be used by each group of students included paper sheets of different weights, pail of water, rulers, scissors, unit cubes, 1cm grid paper, calculators, scotch tape and a digital weighing scale. Each group of students brought their own notebook computer, and internet services were provided to all of these groups.

In this modelling task, students were expected to make different paper boats that could carry different loads (see Table 16.1). The students had to make a few different designs and find out the maximum loads the boat can carry. The issue of aesthetic values against the maximum load is something that the students had to reconcile with. In the modelling process, the students collected data mainly from measurements which they used in the calculations of area and volume of the paper boat. Mathematical concepts such as variables and relations between variables surfaced. Through the worksheet and the process of modelling, we expected the students to discover that mathematical variables such as volume of the paper boat, the base area

Table 16.1. *Schedule of activities.*

Day of MMO	Scheduled Activities
Day 1 (4 hours)	Discuss goals and task details, make assumptions and plans, draw and make the boats, make records and find averages.
Day 2 (4 hours)	Make boats and carry out trials, take photos with loads, make records and interpretations, prepare for presentation.
Day 3 (2 hours)	Prepare for group presentation.

and the weight of the loads the boat could carry would vary according to the design of the boat.

16.4. Observations and Comments

Day 1 was spent by the students on understanding the real-world problem (Real-World Phenomenon, see Swetz & Hartzler, 1991). After conducting a short ice-breaking activity to get all the students to interact and know each other amongst different groups, the facilitators briefed the students on the modelling task and a set of instruction sheets was given to each group. All groups showed a keen interest in the task — first of all by surfing the net to find information on folding paper to make different kinds of boats, some groups even started folding the paper immediately. Thus, the groups were explicitly instructed and encouraged by the facilitators that they had to plan and discuss within the group before folding their boats. A major role of the facilitators was to keep the pupils focused on the task, the two facilitators provided supervision and guidance whenever the students sought assistance. The worksheet that the students needed to complete, served as a guide for the facilitators to check on the students' progress.

In the afternoon of Day 2, the students started to try out their boats on a pail of water (going towards the Mathematical Model, see Swetz & Hartzler, 1991). They put some 1g cubes on the boat until the boat started to sink. In this process, the students had the chance to compare how different designs of the boat affect the carrying capacity. The students also started the "mathematisation process" by recording their measurements related to their designs on the worksheet and carrying out calculations such as finding the surface area and volume. They were quite pro-active in asking the facilitators on finding the surface area and the volume of air in a bottle and bottle cap as they had not learnt how to find the volume of a cylinder.

Most of the groups also started to design or draw their paper-boat models and organised their findings of the boat on vanguard sheets which they would use for presentation to the teachers and the rest of the groups on Day 3. On Day 3, all the groups were able to complete their task on time and carried out their verbal presentation in front of the facilitators, advisors and the rest of the groups (Mathematical Conclusions, see Swetz & Hartzler, 1991). Each group member was asked to vote for the best design that he or she liked best. Before the presentation, the students had to complete a reflection log.

The students used several mathematical concepts. For example, concepts on fractions were used in several ways, one student said; "Half of my boat is carrying loads, while the other half is for passengers". Fractions were also described in choosing the colours, the dimensions (e.g., the height of the boat should be 1/3 the size of the lamppost, so that the lamppost can be seen taller than the boat). We even observed some groups folding the papers into quarters and eighths (i.e., equivalent fraction) to get their dimensions right, while making their boats. Percentage was used to compare between two boats, one student stated; "Boat 2's strength is only 30% to that of Boat 1", indicating that students were able to use percentages as a form of mathematical comparison. The concepts of area, perimeter and volume were used widely throughout in this boat-making modelling problem. The idea of 'the amount of space' in a boat was needed for the students to design the boats in order to decide on how much space (capacity) is available in the boat to carry loads. Although the students have not learnt the concept of capacity, they could relate the concept of space to volume. The concept of ratio was demonstrated by all the groups when they used the idea of scale to estimate the actual real-size object. In order to identify the best model boat, most groups made 2–3 boats and the students also tabulated the most pertinent data such as area, perimeter and volume with sketches for easier referencing and comparisons.

16.5. Discussion

Through this activity, students were exposed to new knowledge that they had not encountered before, For example, Group B (in fact all other groups also noticed later) was the first to realise that the bigger the base area, the more load the paper boat could carry. From the process of making relevant measurements and calculating the surface areas, the students not only applied what they had learnt from the textbook in a modelling situation, they also noticed the weight of the loads the paper boat could carry was dependent on the contact surface area. So the pupils began to see changes in one measurement affecting another quantity — mathematical relationship between variables have in fact taken place in the young minds.

Some students were observed to be very creative and adaptable by using available materials to increase the carrying capacity of the paper boat by sticking bottle caps at the base of the paper boats. The students connected their knowledge about buoyancy from science to this task. This modelling activity indeed brought out the core ideas not just from mathematics but from students' informal knowledge in science as well.

Members of Group D came from three different schools. Initially, we were rather worried that this group would have a problem working together as a team. However, we observed that one of the group members was very creative and had demonstrated leadership qualities in providing help and ideas to the other two members. As a result, they could create about ten different designs and finally focused on improving the best three. One of their final products was voted the best and the team was chosen to present the model to the audience of students, teachers and advisors on the stage in the auditorium. This is an example of how collaborative group work brings out the best of each team member and thus contributing to the overall success of the team.

The students benefitted in many ways from the modelling experience, such as:

(1) The students learnt new concepts while working on the problem.
(2) There was greater opportunity for the students to be innovative and not be constrained by the limits of the problem-solving task.
(3) Students of varying abilities performed at different levels.
(4) The students had to work collaboratively on the task and thereby developed better social skills. This also allowed them to verbalise their thinking which can be an asset for any teacher in class to help them.
(5) Modelling activities provided students with an effective way to have meaningful experiences with complex systems, as "our children live in a

highly sophisticated world composed of interlocking complex systems" (English, 2008, p. 38).

The above benefits have to be weighed against some of the perceived shortcomings. These shortcomings are not necessarily about the modelling process but rather are implementation issues. In particular, in an examination-oriented system such as Singapore, schools will rarely allocate so much time for an activity for mathematics, although the benefits are obvious. The general feeling that students from Singapore are doing well in international studies, not only in the Trends in International Science and Mathematics Study (TIMSS) but also in PISA, which focuses on problem solving in more realistic contexts, may contribute further to the popular feeling from schools, that the mathematics curriculum based on problem solving need not re-focus on modelling.

16.6. Conclusion

The Singapore mathematics curriculum is centred on problem solving, focusing primarily on the four-step problem-solving approach highlighted by Pólya (1957) and problem solving heuristics (see Schoenfeld, 1985). However, the emphasis has not been on authentic problem-solving tasks. One of the major challenges that mathematics teachers encounter is how to use authentic and easy-to-carry-out modelling activity to engage young students. The recent addition of *Applications and Modelling* in the curriculum has not provided explicit ideas about how to implement this in the classroom. There is also the perennial problem of time constraint together with other factors such as a strong examination focus that may explain why mathematical modelling has not been widely implemented in most Singapore primary schools. However, the successful completion of this activity by all the students did demonstrate that folding to build model as a form of mathematical modelling activity can be handled by Primary 5 students. Indeed, as reported by English (2004), modelling activities can be successfully carried out at the primary level. We believe that with appropriate teachers' facilitation and guidance, upper primary students are able to follow worksheet instructions in carrying out a modelling task.

References

Curriculum Planning and Development Division. (2006). *Mathematics Syllabus: Primary*. Ministry of Education, Singapore.

English, L. (2004). Mathematical Modelling in the Primary School. Retrieved 3 Jan 2011 from http://www.merga.net.au/documents/RP232004.pdf.

English, L. (2008). Introducing complex systems into the mathematics curriculum. *Teaching Children Mathematics*. 15(1), 38–48.

English, L., Lesh, R. & Fennewald, T. (2008). *Future Directions for Problem Solving Research and Curriculum Development*. Paper presented at the 11th International Congress on Mathematical Education, Monterrey, Mexico.

Hedberg, J. G. *et al.* (2005). *Developing the Repertoire of Heuristics for Mathematical Problem Solving*. Technical Report for Project CRP38/03 TSK.

Lesh, R. & Zawojewski, J. S. (2007). Problem solving and modeling. In F. Lester (Ed.), *The Second Handbook of Research on Mathematics Teaching and Learning*. (pp. 763–804). Charlotte, NC: Information Age Publishing.

Lester, F. K. & Kehle, P. E. (2003). From problem solving to modeling: The evolution of thinking about research on complex mathematical activity. In R. A. Lesh & H. M. Doerr (Eds.), *Beyond constructivism: Models and modelling perspective on mathematics problem solving, learning and teaching* (pp. 501–518). Mahwah, NJ: Lawrence Erlbaum Associates.

Ministry of Education, Singapore (2005). *Teach Less, Learn More*. Retrieved July 23, 2010, from the Ministry of Education Web site: http://www.moe.gov.sg/bluesky/print_tllm.htm.

Pólya, G. (1957). *How to Solve it?* (2nd ed.). New York: Doubleday & Co.

Schoenfeld, A. (1985). *Mathematical problem solving*. Orlando, FL: Academic Press Inc.

Swetz, F. & Hartzler, J. S. (Eds) (1991). Mathematical modelling in the secondary classroom. Reston, VA: National Council of Teachers of Mathematics.

Printed in the United States
By Bookmasters